Praise for *Desert Notebooks*

"These are the kind of conversations we need to be having—with ourselves and with others. And the desert seems like the right austere
setting to be having them. These fine essays take a deep tradition in
American writing and extend it into our uncertain and collapsing
present."   —BILL McKIBBEN,
                       author of *Falter* and *The End of Nature*

"Ehrenreich's Mojave is both eternal and despoiled, a measuring rod
for the apocalypse, and proof that nature abides. Progress, he explains
to us, is like one of those strange paved streets in the desert running
through phantom, unbuilt subdivisions. The pavement ends abruptly,
and we find ourselves lost in the furnace-hot badlands of the Present
where time and meaning are twisted into enigmatic and terrifying
forms that recall the end-time visions of cultures vanquished by 'civilization.' This haunting meditation on terminal capitalism and its
unthinkable future clearly establishes its author as one of our greatest
essayists, wholly contemporary with these strange times."
                       —MIKE DAVIS, author of *City of Quartz*

"Ehrenreich creates a beautiful meditation on adapting to future
cataclysm."   —*Publishers Weekly*

"The past few years of an accelerated, increasingly destructive climate crisis have brought a number of books that struggle to respond
accordingly to a crisis of such magnitude; several writers have met
this existential challenge with an equally existential discussion of
the ways that the climate crisis affects our understanding of human
history and time itself. Ben Ehrenreich, a columnist for *The Nation*,
takes this discussion to the American southwest, examining the

intersection of science, mythology, and landscape in the desert, in particular in Joshua Tree and Las Vegas. In these settings, Ehrenreich's book reflects on the ways that the prospect of extinction has affected our understanding of time, and how we use that shift in perspective as we move forward." —CORINNE SEGAL, *Literary Hub*, One of the Most Anticipated Books of the Year

"Ben Ehrenreich walked the deserts of the Occupied Territories for his previous book; in *Desert Notebooks*, he takes us with him into the Mojave—its coyotes, creosote, and Joshua trees. He descends barrancas and canyons, hikes boulder-strewn slopes into labyrinthine stacks of Jorge Luis Borges's great Library, from which he draws out stories from that time 'when animals were people,' narratives by the Chemehuevi, the Serrano, the Mohave, and other desert peoples. These echo in texts by Martin Bernal, Walter Benjamin, the Marquis de Condorcet, and Jakob Böhme's mystical touchstone—*The Signature of All Things*—as well as James Mooney's classic, the Ghost Dance and the Sioux revolt of 1890. Climate change California is burning as Ehrenreich's meditations prismatically refract heat, smoke, and light. *Desert Notebooks* is a book for our time—that is, a time scorched by harsh solar rays, shimmering in searing, phosphorescent prose." —SESSHU FOSTER, author of *ELADATL: A History of the East Los Angeles Dirigible Air Transport Lines*

"The crisis humanity faces is total. It's planetary. It's a crisis in space and also in time. How close are we to the end? Is this land we stand on going to be inhabitable in one hundred years, sixty, forty? In sharply featured, compelling prose—the landscape writing here has the heartbreaking clarity of the experience of desert light—Ben Ehrenreich's stunning *Desert Notebooks* combs through history, literature, myth, physics, and ecology to understand how we got here, and how we might find our way out, into forms of time that are made not of our thralldom to capital and petroleum but of our relationships to each other, to our fellow creatures, to plants and rocks and landscapes, and to the stars and sun and moon overhead. Ben Ehrenreich wants you to join him here, on earth. The thrill of *Desert Notebooks* is that in its lucid pages such a miracle seems almost possible."

—ANTHONY McCANN, author of *Shadowlands*

"It's been a long time since I read anything as exciting and illuminating as Ben Ehrenreich's superb new book, *Desert Notebooks: A Road Map for the End of Time*. Very few writers have addressed the current planetary crisis as powerfully and insightfully as Ehrenreich does. The book is extraordinary as much for the rigor of its thinking as for the manner of its writing; its form both narrates and performs the crisis, while also exploring its antecedents. It is, among other things, a remarkable venture in intellectual history, especially in its juxtaposition of the pre-Columbian mythologies of the Americas with the post-Enlightenment mythologies of progress that remade the continents."

—AMITAV GHOSH

DESERT NOTEBOOKS

A ROAD MAP FOR THE END OF TIME

Ben Ehrenreich

COUNTERPOINT

Berkeley, California

Desert Notebooks

Portions of this book appeared in different form in *The Baffler*.

Grateful acknowledgment is made to Rutgers University Press for excerpts from *Black Athena: The Afroasiatic Roots of Classical Civilization (The Fabrication of Ancient Greece 1785-1985, Volume 1)* and *Black Athena: The Afroasiatic Roots of Classical Civilization (The Linguistic Evidence, Volume 3)* by Martin Bernal (1987, 2006); Malki Museum Press for excerpts from *Mirror and Pattern: George Laird's World of Chemehuevi Mythology* by Carobeth Laird (1969), excerpts from *Encounter with an Angry God: Recollections of My Life with John Peabody Harrington* by Carobeth Laird (1975), and excerpts from *The Chemehuevis* by Carobeth Laird (1976); Copper Canyon Press for excerpt from "Abecedarian Requiring Further Examination of Anglikan Seraphym Subjugation of a Wild Indian Rezervation" from *When My Brother Was an Aztec* by Natalie Diaz (2012); and Contra Mundum Press for excerpts from *Eternity by the Stars: An Astronomical Hypothesis* by Louis Auguste Blanqui, translated by Frank Chouraqui (2013)

Library of Congress Cataloging-in-Publication Data
Names: Ehrenreich, Ben, author.
Title: Desert notebooks : a road map for the end of time / Ben Ehrenreich.
Description: Berkeley : Counterpoint Press, 2020.
Identifiers: LCCN 2019047033 | ISBN 9781640093539 (hardcover)
Subjects: LCSH: Desert ecology—Mojave Desert. | Climatic changes—Mojave
    Desert.
Classification: LCC QH541.5.D4 E37 2020 | DDC 577.5409794/95—dc23
LC record available at https://lccn.loc.gov/2019047033

Jacket design by Nicole Caputo
Book design by Jordan Koluch

COUNTERPOINT
2560 Ninth Street, Suite 318
Berkeley, CA 94710
www.counterpointpress.com

Printed in the United States of America

10 9 8 7 6 5 4 3 2 1

For L.
and the new arrival

The only door out is the door in!

—GEORGE MacDONALD

Contents

PART ONE

JOSHUA
TREE

Of an evening the owls come out.

—ARTHUR BERNARD COOK

In the third week of November, one year and six days after the election of the Rhino, I went for a walk with two friends. It was late afternoon. The light was soft, the shadows long. We parked the car on a dirt road about a mile from my house, shimmied under the wire that marked the boundary of Joshua Tree National Park, and walked up through a wide sand wash that I had hiked many times before. As it ascended, the wash would become a canyon: walls of lumpy, reddish stone would rise to the east and west, narrowing as the canyon climbed south into the park. We were following in reverse the path the rain had carved over the centuries as it trickled and sometimes raged down from the rocky hills into the flat, sandy basin below.

Not far from there, in late July, a young couple had gone missing. They were kids from the suburbs, but even if you know the desert it isn't hard to lose your bearings. Canyons fork and twist. The landscape plays tricks on the eyes. The light shifts and familiar terrain becomes suddenly alien. The summer had been a mild one, but most days it was still above 105. Search parties and helicopters scoured the area for weeks. The missing couple came up in every conversation I had in town: Maybe coyotes had already scattered their bones, or they had been abducted by some sinister stranger. Perhaps they had simply wanted to disappear.

In mid-October, searchers found them a couple of miles from my house, and maybe a mile from the wash in which my friends and I were hiking. The young woman's father led the group that found them. The corpses, the newspapers did not neglect to report, were intertwined, embracing even in death. A few days later the authorities revealed that they had found a pistol at the scene: the young man had shot the young woman before turning the gun on himself. Police didn't believe there was any malice in it. The pair appeared to have gotten lost and, having run out of food and water, chose to avoid a slower death. The fact that they had brought a handgun on a day hike was apparently so normal that few of the news reports considered it worth highlighting.

But the summer had passed, the monsoons had poured down in September, and though no rain had fallen since, the senna and brittlebush were still in bloom, smearing the sides of the wash a brilliant yellow. I don't remember what we were talking about—maybe Steve Bannon or the lost hikers or Roy Moore banned from the mall, or the elusive scent of the desert willows that thicketed the floor of the wash—when K., walking ahead of A. and me, stopped. She pronounced a single word: "Owls."

They took to the air in a sudden rustling burst, and then went silent. I barely glimpsed the first one: a flash of wide, white wings as it glided by above us, too big a thing to be so quiet. It soared off in a broad arc and disappeared behind a hill to the west. The second one, though, passed low enough that for an instant I could see its flat, tawny face, the mottled white and brown plumage of its belly, those bright, alien eyes. It circled once and flew out of sight to the east.

Eventually we breathed. With all their circling and swooping, K. thought maybe there had been three of them, but I was fairly sure there were just two. We kept walking, the wash narrowing as we went until we had to scramble over boulders to proceed. We turned a bend. The owls were there, perched on a rock. They saw us first and flew off up the canyon. Again they separated, one arcing right, the other left. We thought that was it and picked up the conversation again. I know at some point we talked about Lebanon and Saudi Arabia, Saad Hariri's strange flight to Riyadh, Jared Kushner's visit the week before. All of that had filled me with a panic that lasted for days, the contours of the next global conflict revealing themselves, requiring only the smallest flame. Who would play the role of the archduke this time? Who would kill him? K. stopped again. The owls had roosted in the rocks ahead of us, as if they were waiting for us there. They flew off and again we watched in silence.

So it went. We scrambled on, following the canyon as it twisted left or right, expecting to see the owls at every bend. Every hundred yards or so we caught up with them and everything we had been saying felt suddenly impertinent, and we fell silent until they flew off and then walked on until we caught up with them again. We talked more quietly now, still surveying the crises of the day, pausing to admire a paper-bag bush in unlikely late-autumn bloom or a particularly bold and healthy cholla. And then the owls shut us up again. We saw them five times in all, maybe six, before they soared off into some more distant canyon and disappeared for good. I knew that we had been annoying them, that they were only trying to avoid us, and it's foolish, I know, but this is what humans do—we turn the world into a story and put ourselves at the center of the plot—and I found

it hard not to imagine, or to want to believe, that they had been leading us onward all along, that they were trying to tell us something,
or to show us a path, one that led deeper into the wilderness, away
from the highway, away from the car.

Before we said goodbye that night, in the parking lot of the
town's one Indian restaurant, the conversation turned to writing. A.
and K. are both writers. It was getting harder, we agreed, to muster faith in any of it, to care at all about lit-world battles that had
once seemed so important. Or even, in the face of real, planetary
disaster—glaciers melting, oceans rising, droughts and fires and famines and floods—to care about something we had once confidently
called literature. No matter how pointless things may have felt at any
given moment, A. said, you could always tell yourself that you were
taking part in a conversation, an exchange that stretched back into
the immeasurable past and on into a future that you couldn't yet
imagine. That was the conceit. Not progress but continuity, at least.
You could tell yourself that it was the conversation that mattered,
this stream of voices flowing through the centuries, this ancient,
almost sacred thing that is bigger and deeper than any of us alone.
But what if it's going to end soon? What if someone in a generation,
perhaps two, will write the very last word? What if the future does
not include enough human beings to keep the conversation going?
What if it drifts off like a party at the end of the night, with only a
few drunks left mumbling in the corners? What if the humans who
remain are too busy surviving to tend to the books and the servers?
What if literacy has a horizon, and it's near? Isn't it all just noise then?

I should add that we were laughing, or smiling, at least. We were
still high from the walk and it felt good to say these things aloud. The
astounding vanity of it, I added, had never felt clearer, this hope that
someone in a hundred years would hear you, that you might be able
to give that person something, just like all the times you had been

lifted and redeemed by the whispers of the dead rustling through the pages of books. How painful and absurd, this fantasy that your own labors might in turn be redeemed by strangers centuries and perhaps continents away who would need to hear what you had to whisper, this delusion that you were doing anything other than babbling because you like the sounds it makes, like a child blowing bubbles into milk. But without those strangers waiting for you, what is the point? Even if *The New York Times* loves you and everyone reads your books today and tomorrow and even next summer, what is any of it worth? Gossip squeaked between lemmings racing for the cliffs. Why bother to write when there will be no one left to read?

Really, I mean it when I say that we were smiling. We were talking about the end of time and the increasingly probable destruction of everything we knew and loved. We didn't relish any of it, but in the context of the walk we had just taken, time took a different shape. The desert enforces its own perspective. It shrinks you and puts eternity in the foreground. If you're open to it, and don't mind a diminished role in this drama, it insists, quietly, on the surging beauty of all things and non-things living and dead and not-formally-alive.

I felt an unfamiliar gladness, soft and pressing, bubbling up. I've thought about it many times in the months that have passed since then: the strange, buzzing joy I felt standing in that parking lot saying goodbye and then driving home alone. Even at the time it felt crazy, like I really was high, though I was entirely sober. It was as if I knew—though I couldn't have known—that I was stepping onto the path that these pages record, as if the joy of discovery preceded the exploration and I were grateful for a journey that I had not yet undertaken, that I didn't even know I was on. I wouldn't start writing until at least a week later, and when I did I had no idea it would become this book. I didn't intend to write a book at all, much less to wage a battle against time, or at least against a certain conception of it, the one that still rules most of our lives and determines how we live them, how we conceive of what has passed before us and of the futures it might still be possible to build.

But that is what I did. That is where those owls would lead me. To fight against that notion of time, I would have to understand how it came to be shaped the way it is, and why we experience it as we do. I would have to ask what histories had to be erased and what new narratives invented for time to rule our lives this way. To figure out, if I could, how those omissions and accretions led us to precisely this perilous moment, in which everything, time included, appears to be on the verge of collapse.

When I did start writing, all I wanted was to remember the owls. I wanted to pin them down like any other memory, so that they wouldn't fade too quickly. One day, if it occurred to me, I wanted to be able to read back and remember what it had felt like: the uncanny beauty of their flight, those late-autumn flowers, the violet light of dusk. But they didn't let me. They wouldn't stop flying. They disappeared behind the rocks and kept reappearing again and again. Weeks after I had left them, they led me to the Maya realm of the dead—you'll see it soon enough—and from there to Hegel and Athena, and to the people who lived where I lived before I arrived there. I won't tell you the rest but I kept following them because I was trying to understand not just time but writing too, and I realized that time and writing are inseparable. Writing extends us in time. It tries to. So that things won't fade too quickly. And by writing I mean something more basic than what gets called literature: the act of inscribing, typing, scribbling, carving, or painting pictographs or glyphs or letters just like these, lines and arcs and loops that stand in for sounds and combine to form words capable of preserving thoughts, ideas, memories, impressions, histories, myths, all the immaterial substance of a culture, its battles over its own past and its present, and its battles over time, and over what it will fight to become.

In any case, I couldn't have known, but there it is: somehow I knew, and I felt happy. Some part of me understood, and didn't know how to tell the rest of me. Sometimes time moves like that, not straight but sideways, backward even, and, like the owls, in silence, in broad and looping arcs.

2.

Yesterday I came upon an article about something called "marine ice cliff instability." The idea being that as ocean temperatures rise and icebergs break away from the glaciers that cover West Antarctica, they reveal higher and higher cliffs of ice. If the cliffs reach a certain height, the ice will no longer be able to support its own weight and will begin to crumble off in giant shards. Enormous, skyscraper-sized icicles will splash into the sea, each one rendering the cliff behind it taller and more unstable and prone to collapse. In other words, it could all go very suddenly.

"The destruction would be unstoppable," the article pronounced. This could happen before the century ends. In the next twenty years even. It could mean that in our children's lifetimes, if not our own, the oceans would very swiftly rise eleven feet or more, nearly four times as much as previously projected. Mumbai would be inundated. So would Hong Kong, Shanghai, New Orleans, Jakarta, Lagos, south Florida, and Bangladesh. New York and London would not fare well. Not millions but hundreds of millions of people would be displaced.

I read an article like this almost every week. I don't look for them. They show up on my Twitter feed in the morning over coffee, with the day's eructations from the Rhino and funny alpaca GIFs and the latest in police killings. From last week, November 15: climate scientists forecast that temperatures are likely to rise 3.4 degrees Celsius by the end of the century, more than twice the 1.5-degree target agreed to in Paris by every country in the world except the United States. From November 2: a new report—I have come to fear the words "a new report"— predicts climate change will push tens of millions of people from their homes in just the next decade, "creating the biggest refugee crisis the world has ever seen." On October 30 there was another new report: global emissions of carbon dioxide, which had appeared to be tapering off, leaped in 2016 by more than 50 percent over the previous year, reaching a level not observed since the mid-Pliocene era, approximately three million years ago. Whatever happens, there is no reason to doubt that human civilization, and all life on the planet, will be radically reshaped. On October 13: French scientists announce that thousands of penguins have starved to death in Antarctica. In a colony of forty thousand Adélie penguins, only two chicks survived.

I finished my coffee. I took a shower, got dressed, and thought about those owls.

I remember, when I was a kid, staring at road maps, the kind you bought at gas stations and carried in the glove box, and that were, for me at least, impossible to properly refold. I remember looking at all those intersecting lines representing roads laid over and carved through the earth, dirt tracks and superhighways, the insolent grids of the cities. I wanted to follow them all to the end. I remember thinking that if you could get hold of all the maps for the entire country, or even the hemisphere, and spread them out side by side, it would be obvious that every road leads to every other road, that everything is connected. The dull suburban lane on which I lived would carry me eventually to rocky paths in Patagonia and the rutted logging roads that cross Alaska. There were dead ends, of course, lots of them, but assuming you were free to backtrack, it was impossible, really, to get lost. You could follow any road in any direction and eventually, by however circuitous a path, get where you needed to go. Oceans notwithstanding.

I don't remember talking to anyone about this. As a child you learn to guard your thoughts, to hold close to ideas that seemed simple and self-evident and that you knew adults would scoff at. What counted as education seemed to mainly involve learning to walk in single file and otherwise keep quiet. School meant grown-ups telling you that things had to be done in a certain way, and in no other, that however many obvious and inviting paths might lead from one point to another, only one of them was right. The rest might as well not exist at all. To do well, to earn praise, you had to learn not to see them anymore.

I've had some time to think it over and I'm convinced I was correct. For decades we have been told that political maturity meant accepting that there were no alternatives to the world in which we lived, that no deviation was possible from the path that we were on. That economic growth was limitless and democracy would advance alongside it, and prosperity, equality, freedom, and endless high-tech toys. That to question this, to strive to imagine any radically different way of going about things, was a childish and even dangerous endeavor. That our society had evolved over the millennia via the straightest route available—the only one, at that—from a pitiable primitive infancy to the heights of rational civilized society, and that our only option was to continue to climb the same path. That was the story, and somewhere along the line most of us began to believe it. It helped us to forget that there were always other roads, other ways to see things, other stories, other routes. We didn't see, most of us, that the path we were on would lead us here, into this cul-de-sac. Now the asphalt is melting, and falling away beneath our feet.

We have no choice but to scramble to retrace our steps and to try, in a hurry now, to imagine things differently: other worlds, other ways of thinking, living, seeing. Other ways of writing, and of reading. By "we" here I mean nearly all of us, no matter where we started, or where we've ended up. This means telling other stories, and listening to them—perhaps especially to the ones discarded long ago, and to the ones told by people whose paths collided with that of the society in which we live and who did not make it through that encounter intact. It's not because anyone else was so much better or smarter that we should listen—though some of them may have been—but because they are not us, and we need all the help we can get to become something else.

There's not much time, but remember that all roads are connected. If we hurry, we could follow almost any route, really any road we want, so long as we pay enough attention along the way. If you saw a crow fly by the window as you were reading just now, you could follow it and see where you end up. If you saw a plastic bag blow past you on the wind, you could follow that too. But I live in the desert now, and I saw owls, so I will follow them.

ॐ

As far as I can figure they were short-eared owls. The drawings in my copy of the *Sibley Field Guide to the Birds of Western North America* looked just like them. They had the same dark-rimmed eyes and dusky, mottled faces. The internet, however, suggests that no short-eared owls reside anywhere near Joshua Tree, California. But none of the owls that are supposed to be native to the area look like the ones we saw. So maybe they were migrating. Or they were some other kind of owl. Or I dreamed it all. Or the owls dreamed me.

Later A. told me that he saw them again in that same wash and that he was confident they were barn owls. Maybe so, but when they first flew over our heads and we were standing, jaws at our navels, gawking, among the willows in that wash, he had mentioned another kind of owl. A. lived for a while in Guatemala and has read almost everything. His first thought that afternoon was of the *Popol Vuh*, the so-called "Council Book," which records the creation tale of the K'iche' Maya. "Remember the owls in the *Popol Vuh*?" he said, still grinning. "Messengers from the lords of the dead."

I had skimmed it years before, or thought I had, but I didn't remember any owls. Later I looked them up, and kept looking things up. I was curious, though that may be too polite a word for the hunger that I felt. I wanted to know where the owls would take me. I started with the *Popol Vuh*. I read it twice, in two translations, and I kept reading everything I could, following whatever paths opened themselves up to me. The *Popol Vuh* mentions four kinds of owls. Or better put, four owls. They served as messengers for the Lords of Xibalba, the Maya underworld. The Lords were a nasty bunch, devoted to the torment of humans. There were two whose work it was "to make men swell and make pus gush forth from their legs," two who made men "waste away until they were nothing but skin and bone and they died," two who caused men to suddenly begin vomiting blood until they died on the road as they walked.

Xibalba was not quite Hades or Hell, but an entire dimension of terror with its own detailed geography of punishment: awful mountains and rivers of blood; a house through which a cold wind blew, "where everybody shivered"; another filled with giant, murderous bats squeaking and screaming and flying frantically about; another teeming with knives that could dart through the air without a hand to hold them; a "house of gloom" in which there was only darkness. This was not metaphor. Xibalba was a real place, somewhere in the West. You could get there through a cave or a *cenote*, one of the underground springs that dot the Yucatán

Peninsula, or by following the "Black Road," the dark cleft at the center of the Milky Way, all the way to the horizon.

But the owls. They make their first appearance in the *Popol Vuh* when the Lords of Xibalba become annoyed with two brothers named One-Hunahpu and Seven-Hunahpu, who, like many young men, did not like to do anything but throw dice and play ball. One day the brothers were playing ball on the road to Xibalba. They were making lots of noise. The Lords, furious at this show of disrespect, dispatched the owls with a message for them, at once a summons and an invitation: "They must come here to play ball with us so that they shall make us happy," the Lords told the owls.

The brothers obeyed. What else could they do? The Lords of Xibalba amused themselves with a series of unpleasant tricks, like inviting the brothers to sit on a bench of burning-hot stone—they found this quite hilarious. Then they murdered them. Before they buried them, though, they took One-Hunahpu's head and hung it from a tree that had until that day always been barren. Instantly the tree was heavy with fruit. The Lords of Xibalba issued an edict: no one should ever eat from the tree or even sit beneath it.

Such edicts are inevitably defied. Word got around. A young girl named Xquic, which means Blood Moon, heard about the miraculous tree. Imagining that its fruit must be impossibly sweet, she decided to seek it out, to taste it. They are rare, but there are people like Xquic everywhere, fortunately, people who defy the edicts of the powerful, whose curiosity rejects all constraints. When at last she found the tree, and stood beneath it, the skull of One-Hunahpu, which still hung from its branches, asked her what she wanted.

"These round things hanging from the branches are nothing but skulls," it said. "You don't really want those?"

"Yes," she said. "I do."

"Okay," said the skull. "Just hold out your hand."

When she reached out her hand, the skull dribbled spit into her palm, and Blood Moon became pregnant. With twins, of course.

This is an odd story, I know. Is it any odder, though, than the one about the god who loves us absolutely and out of incomprehensible divine love gives us not only his despised half-mortal son, whom we murder and reject, but a strange and mystical substance known as "free will," which condemns us, by and large, to reject our creator again and again and suffer unending sorrows? Or the one in which without realizing it humankind has been riding for the last few thousand years on a clunky sort of spaceship called *progress* that is taking us—some of us, anyway—to a better place, in which the miracle of reason yields universal happiness, comfort, and health?

Seeing that she was pregnant, Blood Moon's father complained to the Lords of Xibalba that his daughter had been disgraced. They instructed him to question her and, if she refused to answer honestly, to sacrifice her to them. So when Blood Moon protested to her father that she had never slept with anyone, he brought in the owls. Kill her, he ordered them, and bring her heart to the Lords of the Dead.

The owls carried Blood Moon off to Xibalba. As they flew, the girl was able to convince them that their orders were unjust. They would like to help, they told her, but what could they do? They had been ordered to take her heart to the Lords.

"But my heart does not belong to them," Blood Moon said. "Neither is your home here, nor must you let them force you to kill men." She was persuasive, and she was right, so the owls, at her urging, rebelled. They cut into the trunk of a tree with sap that ran as red as blood. They shaped the sap into a ball, which they pinched and formed until it resembled a heart. The owls brought this heart to the Lords of Xibalba, who were satisfied that it belonged to Blood Moon. Then they flew up out of the abyss, abandoning their masters to join Blood Moon again. She would give birth to two boys, Hunahpu and Xbalanque, the trickster hero twins of Maya lore who, after many trials, would humiliate the Lords of Xibalba, avenge their father, and topple the dominion of the dead.

This is a long way of saying that sentences are not always final. Messengers do not always obey. Owls can be dispatched with one message and return with another. A single message, no matter how apparently unambiguous, can mean more than one thing. I'm counting on it.

≋

I almost forgot the letter. I didn't see it on Twitter, or in the papers, or on the news. I only found it just now while googling to check the facts in the section about the ice cliffs. On November 13, more than fifteen thousand scientists signed a "Warning to Humanity" noting the rise in carbon emissions, the depletion of freshwater resources, the growing "dead zones" in the seas, the destruction of the forests, and the unleashing of a "mass extinction event . . . wherein many current life forms could be annihilated or at least committed to extinction by the end of this century."

The scientists urged political leaders to act and made some suggestions for the sort of sustainable management of the earth's resources that might save us if enacted in time. Given the urgency of their warning, their proposals felt quite modest, perhaps not entirely adequate to the task: "prioritizing" the establishment of vast reserves to restore forests and native plants, "rewilding" swathes of the planet, "promoting" a shift to a largely plant-based diet, reducing human fertility by ensuring access to birth control, "massively adopting" renewable energy sources while cutting back subsidies to the fossil fuel industry, "revising" the global economy in order "to reduce wealth inequality," etc. With one exception—"increasing

outdoor nature education for children, as well as the overall engagement of society in the appreciation of nature"—every single one of
their suggestions was, in the pragmatic terms that politicians favor,
inconceivable. Given budgetary restrictions on nonmilitary expen-
ditures, even that latter prescription would be a hard sell.

Until the Rhino* upended everything, we had grown accustomed to politicians telling us that we must be practical and modest,
that we should not expect too much. (Don't make noise on the road
to Xibalba.) Barack Obama's favorite line: "The perfect is the enemy
of the good." No matter that the good slipped off along the way and
was replaced by the shitty a long time since, this was the mantra
of ruling bureaucracies around the world for decades: that a rational politics can provide only incremental change; that any attempt
to ask for more would be divisive and ultimately disastrous; that
we should not fret because, to borrow Obama's other favorite line,
which he borrowed from Martin Luther King, the arc of history
bends toward justice. We could have faith in progress if no other
god. Time has a shape and a direction. We might not be able to see
the arc in its entirety, and we should not be so bold and foolish as
to hurry it along, to demand justice, or much of anything, but we
should know, and be comforted, that however it might seem, step by
step, compromise by compromise, things are getting better.

So the dogma goes. Or so it went. But those compromises, that
refusal to make any demands that might upset the system—and irritate the Lords who profit from it—led us to this precipice, and the cold

---

* I owe an apology, I realize, to that otherwise marvelous beast, the rhinoceros, which shares none of the president's malevolence, deceitfulness, and dizzying weakness of character, and does not deserve to be tarred by this analogy.
But however noble it may be, the rhinoceros is not a delicate creature, and the
Rhino, whatever he may intend, only ever wrecks things.

winds spiraling up from Xibalba. Decades of scrupulous and unrelenting pragmatism carried us here. The minimum necessary for survival now counts as madness. The courses of action still deemed practical will usher us straight down the path that leads to our own deaths.

The owls flew away. They didn't fly straight but swooped in long arcs, hidden by the dusk, suggesting new paths and retracing old ones all at once. Pragmatism reeks. I want out. A way out that is first of all a way in.

⁊

In the end, Hunahpu and Xbalanque overcame the Lords of Xibalba by repeatedly faking their own deaths. Or, better put, by actually dying and then coming back to life. They jumped into a bonfire, burned to death, had their bones ground to powder and scattered in a river. But they had it all planned out: at the bottom of the river "they changed back into handsome boys." After five days, disguised as magicians, they returned to Xibalba and worked many miracles for the pleasure of the Lords. They burned houses and made them whole again. They killed a man, cut out his heart in sacrifice, and brought him back to life. They killed each other, threw each other's hearts on the grass, and then returned to life. The Lords of Xibalba couldn't get enough. "Now do us!" they shouted. "Cut us into pieces, one by one!" And so they did.

Death is not always an end. The *Popol Vuh* may be one of the only texts in existence that records its own destruction. The version that has been passed down to us was translated and transcribed in the early 1700s by a Dominican monk named Francisco Ximénez, who recorded the text in twin columns, one in Spanish and the other in the language of the K'iche' Maya, rendered phonetically in Roman letters. The source the monk copied has never been found, but archaeologists believe it was likely an older manuscript in phonetically rendered K'iche' that can be dated with some confidence to the mid-1500s. Among the clues that make that dating possible is the fact that the surviving text refers to the city of Santa Cruz, the name bestowed by the Spanish in 1539 to the former K'iche' capital of Utatlán, which had been leveled fifteen years earlier by the conquistador Pedro de Alvarado, the most trusted, the most charming, and the handsomest of Hernán Cortés's lieutenants.

Realizing after two profound defeats that they could not beat him on the battlefield, the two kings of the K'iche' had sent a messenger to Alvarado. They had invited him to Utatlán to discuss the terms of their surrender. Perhaps correctly, Alvarado feared a trap. He refused to leave the plains beneath the capital, which was as much a fortress as it was a city, high in the hills above a deep ravine, with stone walls and narrow streets. Instead he invited the kings to visit him. On March 7, 1524, they complied. Alvarado later testified: "As I knew they bore ill will to His Majesty, and for the tranquility and well-being of that land, I burned them and I ordered the city burned and razed to its foundations."

The manuscript from which Francisco Ximénez copied the *Popol Vuh* was likely transcribed in the years that immediately followed this holocaust, copied out hastily and hidden from the invaders. It was a task performed with the awful knowledge that nearly everything was lost. The book ends by mourning its own disappearance: "And such was the existence of the K'iche', which can no longer be seen anywhere, because the original book [the *Popol Vuh*], which the kings had in ancient times, has disappeared. So it is, then, that everything about the K'iche', which is now called Santa Cruz, has come to an end."

≈

We are not the first people to believe we are living at the end of time. Far from it. The K'iche' understood that their world was ending. At some point between the arrival of Europeans in the late 1400s and the close of the nineteenth century, so did most of the people who had been living in the Western Hemisphere. For the Aztecs and the K'iche' and the Inca and many others it came quickly, and cataclysmically. For the inhabitants of the Great Plains of North America and the desert Southwest, where I now live, Armageddon would be slower to catch up. That apocalypse is always with us: all the joy that I take from this land has been contingent on the destruction of someone else's world.

It is not always so violent. The Maya had experienced a previous collapse, in the ninth century, when the sophisticated lowland cities of their empire were precipitously abandoned without any aid from Spaniards, firearms, or smallpox. Drought, deforestation, soil degradation, and, perhaps, the arrogance of an unresponsive elite, were enough to do it that time. Which is to say, the same things that will likely do us in: the greed and blindness of the few, the hungers of the many, a fatal inattention to the fragile web of life on which our existence here depends. The first few meters of the surface of the planet are littered with the remains of dead civilizations, people for whom the world has ended and the circle of time has closed. Why should we be special? Bones, tissue, hair quickly become soil, but metal, stone, and baked clay can last a few thousand years, long enough to keep the archaeologists in grant money for a little while longer.

There have been plenty also who were too hasty to conclude that their world was ending, countless chiliastic sects and prophets proved premature by the failure of the rapture to arrive. Recall the Baptist preacher William Miller, whose calendrical calculations and close readings of the Book of Daniel led him to predict—and his many thousands of followers from New England to Australia to believe—that at some point between March 21, 1843, and the same date of the following year history would end, and Christ would return. When that twenty-first of March passed without incident, he revised his calculations, nudging the date forward by one lunar month, to April 18. Still earthbound on the morning of the nineteenth, Miller realized that he had fudged the math again. Christ would come in autumn, he was positive this time, on October 22. But Christ did not come, not that fall and not any season since, and in the despair and disappointment that fell upon the thousands of the faithful, a world really did come to an end.

I, on the other hand, will be thrilled if I am wrong about everything. But then I don't think Christ is coming, or the Messiah or the Mahdi or the Martians. It's worse than that: no one can save us but ourselves.

≋

Since I am writing here in part about writing, it is worth adding that Hunahpu and Xbalanque had two older half brothers, Hunbatz and Hunchouen, who were born before their ball-playing father took his unfortunate trip to Xibalba, from which he did not return. Those other sons are regarded as the patrons of artists and writers. They were jealous of their younger half brothers, the twins, and neglected and abused them, so the twins turned them both into monkeys.

≋

A few days ago, scrolling down my Instagram feed, I paused on a post from the NASA account. The image was a computer simulation of two black holes colliding in space. In September, astronomers were able for the first time to record the shape of gravitational waves rippling through space-time. They arrived as an almost undetectable wobble in an otherwise unremarkable transmission, the quivering, 1.8-billion-year-old remnant of two black holes colliding. Someone had animated a video depicting two black holes circling one another like boxers looking for a gap in each other's defenses, dragging the stars around them into furious orbit until they eventually and very suddenly combined into a single black hole the size of fifty-three suns. The Instagram post was a still from that video. It looked like the face of an owl.

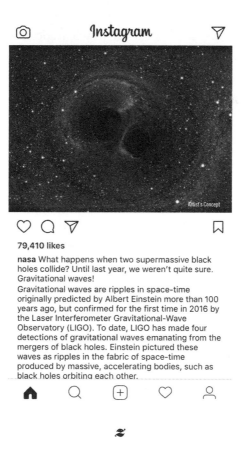

When L. and I first moved to the desert, we set ourselves the task of learning the stars. Except for a few periods of a month or two in a rented cabin here, I had lived in cities all my life and never managed to learn anything beyond the most easily identifiable constellations: Orion, the Big Dipper, the Pleiades.

L. is more disciplined about these things than I am, so every night we bundled up—it was early winter then—stood outside the house, and studied a new sector of the sky. We extended the diagonal line formed by Orion's belt down to Sirius, in Canis Major, and up to Aldebaran, the brightest star in Taurus. (John Berger: "One can lie on the ground and look up at the almost infinite number of stars in the night sky, but in order to tell stories about those stars they need to be seen as constellations, the invisible lines which can connect them need to be assumed.") We learned to imagine a line between Orion's right foot, Rigel, and his left shoulder, Betelgeuse, and to follow it out to Gemini, the twins—not Hunahpu and Xbalanque but Castor and Pollux, both hatched from the same shiny egg after Zeus, taking the form of a swan, raped their mother Leda.

The weeks passed and the winter progressed. We traced a new line between the two stars on the inside of the Big Dipper's bowl and followed it to a great backward question mark that had not been visible a month before and that formed the mane and shoulder of Leo. Later, in the spring, we watched Scorpius and Sagittarius rise and set and saw how the constellations of the zodiac—and the planets and the moon—travel along an established path, a sort of highway across the sky, the ecliptic, which sways a little as the year progresses, and then sways back again. We read what we could and watched videos on YouTube so that we could better visualize and understand what these movements meant, how they reflected the shifting position of the Earth in the universe, its tilted orbit around the star we call the sun.

I remember standing in front of the house and almost falling down in a moment of dizzying comprehension, staring at the Milky Way or the polestar and understanding with my body as much as my mind where we were in the universe and how and where we were moving. I felt like I'd been punched. More than with any political or philosophical revelation, the entirety of my perspective on existence—which, despite all my convictions and everything I understood in the abstract, was nonetheless centered on the earth, and for the most part on this particular North American landmass, and on this minuscule body and the tiny and petty radius that extends from my eyes and thoughts and emotions—all of it shattered. I am only exaggerating slightly. I had my passport number memorized, my Social Security number and my street address, my PO box and zip code, but I had never known where I lived. Or where we are going. Whenever I arrive in a strange city I learn the basic layout of the streets as quickly as I can so that I don't do something stupid or get lost, but for my entire life I had somehow neglected to take this most fundamental step, one that humans had been taking for millennia.

I began to understand, as the Maya did, and the Greeks and Egyptians and Somalis and Indians and Sumerians and Chinese did, that time and space are inseparable. The sky is a clock, and a calendar is also a map. To know a date and a time is to know the positions of the planets and the stars, their relation to one another and to us. To know where the stars are is to know what time it is, what day and what year. Time is not an independent vector that pushes on, stubborn and cocksure, taking us to a place called the future. It lives in our bodies and in the stars, in the mountains that rise up from the sea floors, in the wind and rain that wash the mountains back into the sea. Everything moves. Mountains and oceans as well as stars. Orion disappears beneath the horizon in the spring and rises again in autumn, though I learned sleeping outside in the hammock on hot nights that even in August, he appears before dawn. If you stay up late enough, you'll see the next season's stars cycling past.

Every method I can think of that human beings have devised for representing time depends on displacements in space. The hands of a clock, the swing of a pendulum, the shifting shadow of a sundial, the vibrations of the electrons in a cesium atom. Time is motion. It favors circles, spirals, and ellipses, but it also does not mind explosions, supernovas, catastrophic disruptions that appear to halt the rhythmic flow. It is worth considering too, though, that those disruptions and all their violence are part of some larger cycle, one so immense that we are not equipped to trace the path of its orbit.

If only we could stay up late enough and see what rises next.

I read in the paper this morning that the Rhino, in the middle of a ceremony to honor two Navajo veterans of the Second World War, could not resist a jibe at a senator who had claimed native ancestry, calling her Pocahontas, as he has on several occasions before. This time no one laughed. At least some of the people gathered in the Oval Office likely knew that the historical Pocahontas was the daughter of the Mattaponi chief Wahunsenaca, whose people had the misfortune of living a few miles from the site chosen by the English for the settlement of Jamestown. When she was about fifteen, Pocahontas was kidnapped by a man named Samuel Argall, who later claimed that he did not abduct her but traded her fair and square for a small copper kettle. According to Mattaponi oral history, if not the English texts, she was raped while in the custody of the English and gave birth to a son. He would be named Thomas. Later she was married to one John Rolfe and given the Christian name Rebecca. Rolfe brought his young wife to England. She sailed on the same ship that two years later would carry the first enslaved Africans to the Virginia colony. Rebecca Rolfe, born Pocahontas, died at the age of twenty, far from home, in the English town of Gravesend.

Except for its extensive use in myth and marketing, there is little that is unique about Pocahontas's story. She was far from the only indigenous woman to be so abused. But alongside the many efforts preserved in print and on celluloid to twist the tragedies that befell her into the honeyed pap required for exonerative nation-building, another history survived. Without books and without paper, in the long shadow of a genocide, another narrative persisted, from mouth to ear: they took her and they raped her; she did not love him or any of them.

It's nothing to be smug about, but it is possible that we are no worse than the rest. And by "we" this time I mean modern-day Americans with our SUVs and our HDTVs, our outlet malls and our prison archipelago, our active shooter alerts and fracking-induced earthquakes, our escape rooms and tent cities, our forward operating bases and our concentration camps for the immigrant poor. We are not that much worse anyway, perhaps. The Maya did not shy away from human sacrifice either, and the Aztecs made an industry of it. People fought wars and slaughtered and mutilated and abused each other throughout the Americas long before any Europeans arrived here, as humans have in every other place. Most states from ancient Babylon onward have been even more brutally unequal than our own, and slavery has, almost universally, been coeval with what we like to call civilization. Humans living before the rise of organized states and even before the advent of agriculture left a legacy of enormous devastation: the ongoing, sixth wave of extinctions did not begin over the last couple of centuries but thousands of years earlier, with the extermination of species after species of large mammals, most likely thanks to human recklessness, on nearly every continent. The spread of *Homo sapiens* across the Western Hemisphere coincided with such a vast decimation of large herbivores—think mammoths and camels and oversize sloths—that some scientists have speculated that the resulting loss in atmospheric methane (herbivores do a lot of belching and farting) was enough to alter the climate, causing a drop in temperatures that lasted thirteen hundred years. That's one theory, anyway. More likely it was a meteor colliding with the earth. Worlds end all the time.

All of this is to say that human beings have fucked up before, but never have we fucked up as we are fucking up now. In less than fifty years—not even the heartbeat of a gnat on a planetary scale—we have eliminated 60 percent of the mammals, birds, fish, and reptiles that populated the earth. The insects and amphibians aren't faring any better. As many as a million species are now facing swift extinction. We have transformed, and not for the better, 75 percent of the surface area of the planet, destroyed 85 percent of its wetlands, and befouled, to varying degrees, two-thirds of the volume of the oceans. We could, if you want, sigh and shrug and agree that humans suck, that we are a rogue species, a kind of poorly evolved virus that can't help but kill its host, some bipedal breed of demons, a curse, and that we cannot avoid our fate. Plenty of evidence would support these conclusions.

Surrendering to inevitability can feel pretty good, but it does not get us off this hook. Despite our Paleolithic ancestors' wanton overhunting of mastodons and long-nosed peccaries and flat-headed peccaries and giant beavers and gianter sloths, and despite our species-wide eagerness to mistreat one another more or less everywhere and all the time, no other humans have managed to be destructive on anything close to the scale that we have over the last two centuries and change. (Not even the wing beat of a midge . . .) Or, I should say, on anything close to the scale forged by one relatively small subsection of humankind—mainly Europeans and those, like me, of European descent living in what gets called the Global North, the last two centuries' inheritors of what some folks still hail as Western Civilization. West and North are hardly precise geographic terms here, but you know who I mean: we fucked things up for everyone.

I take no pleasure in this verdict. Self-flagellation is another form of narcissism. But if despair is an indulgence we cannot afford, so is delusion. Among all the shrieking and shouting and fearmongering and warmongering and the mad, panicked bellows of the Rhino, I occasionally hear some talk out there that is not entirely insane. Murmurs and whispers at the margins, calls to unmake the economy that brought us here and, while there is still time, to find some way to build a new one that does not depend on the illusion of eternal, self-sustaining growth, one that is based neither on the massive exploitation of fossil fuels nor on the systematic exploitation of other human beings. This is no small task, I know, but it will be doomed in advance unless we also work to dismantle our delusions: the flattering stories we tell ourselves about ourselves, the myths that structure our existence so seamlessly that we don't know they're myths, the ones that carried us here, and with us everyone, as surely as all those coal- and diesel-powered engines did.

z

Actually, there is one kind of clock that does not use space to measure time. The ancient Chinese made "incense clocks," which relied on the consistent rates of combustion of different varieties of incense. The scent changed to mark the passage of the hours.

That's how it is, isn't it? Time has a certain smell.

z

I grew up in the seventies and eighties in a family that lived and breathed politics. Dinner conversation leaped from Falwell to the Sandinistas to the nuclear meltdown at Three Mile Island. Pass the salt and the gravy—you need a napkin there, kid? The height of the Cold War had passed, but I had heard enough adults express alarm that Reagan's cowboy bullshit was going to get us all killed that I did not expect to live to adulthood. I don't remember feeling sad about this, just accepting it, that life had a horizon, and it was close. I read a lot. I asked a lot of questions. I knew about the dread mechanics of nuclear winter and the various stages of radiation poisoning. I knew to squeeze my eyes shut when the blast came because the flash could burn out your retinas. (But wouldn't it burn through your eyelids too?) Manhattan, just thirty miles away, would without question be a target. So would the Grumman plant a few towns over. I spent long afternoons thinking hard about what I would do if I survived the initial impact, whether I should find a way to kill myself or take my chances and live on as a mutant, my skin peeling off in sheets. If I had to, would I be able to eat the dog?

Then it ended. The Soviet Union fell. The Cold War was over. I read J. G. Ballard and every work of apocalyptic fiction I could find. I could rhapsodize if you let me about Octavia Butler and the Strugatsky brothers and Russell Hoban's *Riddley Walker*. I thought Denis Johnson had been going downhill ever since *Fiskadoro*. I wrote stories and novels of my own. Cataclysm loomed in nearly all of them, and a vague sense of guilt. But time had not ended. On it ticked, and if I'm honest I felt lost, agoraphobic. Would it really stretch on forever? Occasional panics ensued, warheads gone missing from the old Soviet stocks, Ebola, dirty bombs planted out there somewhere by CNN's latest villains, a nagging sense that the twentieth century was not finished with us. I kept busy. There was plenty to get upset about and plenty left to fight for, but the fear of full-on planetary catastrophe wandered off for a decade. Maybe two.

It's back, of course, but not like before. In the eighties there was just Reagan and the Russians. Now the danger comes from ever-multiplying fronts, all of them at once. There's the Rhino and the North Koreans, and the Rhino v. Iran. There are the wars in Iraq, Syria, and Yemen in which nearly all the major military powers on the planet spread death by proxy because, for now, they are not ready to face off directly. There is the unlikely romance between Netanyahu and the Saudis, impatient with this by-proxy crap. There are the fascists in European parliaments and the fascists in the streets and the fascists in the White House. There is the vast and lethal infrastructure of Bush and Obama's war on terror waiting for the tiniest excuse to leap to yet another battlefield. And on top of it all, beneath it all, on every surface and every side of everything, there is the already ongoing mass extinction, the melting glaciers, the jet stream stalling out, the droughts and the fires, the gathering storms. Time has a stench.

This morning I checked the news on my phone while brushing my teeth and read that North Korea had tested a missile capable of reaching Washington, D.C. In other words, almost anywhere in the United States. The Rhino's response so far has been restrained, or perhaps just distracted. He spent the morning retweeting videos posted by a leader of an obscure English ultranationalist group. Everyone on Twitter was indignant as always, but I felt only relief. Today, at least, he was too scattered to insult Kim Jong-un.

If time really is ending, if these are end times, maybe it is worth pausing to ask: What is time? How do we understand it? Why do we experience it the way we do? Have that understanding and that experience helped to lead us here, to this precipice and the particularity of this specifically bad smell?

Eventually Hunahpu and Xbalanque got a rat to tell them where their father's ball-playing gear was hidden. It was hanging from the roof beam of their grandmother's home. The rat gnawed through knots and the brothers scooped up the gear. They began to play, as their father and uncle had. The Lords of Xibalba heard them playing and sent a messenger. Not owls this time—they had escaped by then—just a messenger. Their grandmother answered the door. The twins weren't home, she said. The messenger delivered his message. Again: a summons, and an invitation.

Grief froze the grandmother's heart. The last time she had received such a message—an identical message—she had lost her sons forever. She promised to pass the message along and closed the door. She sat there by herself in her home and wept in silence. She mourned her sons, gone all these years, and the grandsons who would soon be lost to her as well. A louse fell from the ceiling and into her lap. She picked it up. She let it crawl in her palm. She watched it for a while, and then she got an idea. She spoke to the louse. She called it "my child" and asked it to pass a message to the twins. She told it what the messenger from Xibalba had said, and the louse hurried off, pleased to have a purpose. But the louse was soon swallowed by a toad, and the toad by a snake, and the snake by a hawk. The grandmother almost got her way.

Not all messages reach their intended audience. Sometimes the messenger is killed, sometimes the audience. Writing in the mid-sixteenth century, the Dominican cleric Bartolomé de las Casas confided that Maya texts were crafted "with such keen and subtle skill that we might say our writings were not an improvement over theirs." I will be bolder and say that no people anywhere have devised a more beautiful system of writing. Nearly all of their texts are lost.

In 1549 the Franciscan monk Diego de Landa first traveled to Mexico. He was a young man of twenty-five. What he saw disturbed him. In his *Relación de las Cosas de Yucatán*, de Landa wrote of Spanish soldiers burning men alive, of seeing women's bodies hanging from trees, their children tied dangling to their feet. De Landa witnessed, he wrote, "unheard-of cruelties, cutting off noses, arms and legs, and women's breasts, and they threw [the Maya] in deep lagoons with gourds bound to their feet; they stabbed infants because they could not march like their mothers, and if they shackled them with collars and they fell ill, they would cut off their heads so as not to have to stop to free them." The Spanish justified these infamies, de Landa wrote, by insisting that they could only hope to dominate such a vast population by inspiring fear incommensurate with their numbers. Perhaps they were right.

About a decade after his arrival in what was then called New Spain, de Landa was charged with bringing the word of God to the Maya of the Yucatán. He had acclimated by then. Or perhaps he had learned to make a distinction between gratuitous slaughter and slaughter for Christ. For the Spanish the Americas were, in Eduardo Galeano's words, "the vast kingdom of the devil." De Landa declared himself grand inquisitor and did his best to stamp out the idolatrous practices of the natives. Francisco de Toral, the first bishop of the Yucatán, would recount that de Landa and the three monks accompanying him bound and whipped the Maya they encountered in order to force them to confess to worshipping idols. If their victims endured one hundred lashes, "they would hang them publicly from the beams of the church by their wrists and attach a great deal of weight to their feet, and they would burn their backs and their bellies ... until they confessed." Many did not survive this procedure.

De Landa imported to the Yucatán another key institution of the Inquisition, the auto-da-fé. He wrote: "These people also used certain characters or letters with which they wrote in their books their ancient affairs and their sciences, and with them ... they understood their affairs and gave them to be taught and understood. We found a great number of books containing these letters, and as there was nothing in them but superstition and the falsehoods of the devil, we burned them all, which they regarded as a wonder and which caused them much affliction." On a single day in July of 1562, de Landa is said to have burned at least twenty-seven codices, the elaborately inscribed fanlike texts that the Maya used as books.

In all of the lands inhabited by the Maya, which stretched from what is now southern Mexico through Guatemala and Belize into the western regions of El Salvador and Honduras, a total of four codices survived this and other conflagrations. An entire literature reduced to four crumbling volumes. In many places more than 90 percent of the population disappeared within a few years of the conquest, annihilated by a combination of depredation and disease. With the destruction of the codices and temples—the painted and sculpted walls of which were also vivid texts—and the decimation of such a vast portion of the populace, literacy reached its horizon. Within a few generations, the Maya who survived could no longer read the texts written with such care by their ancestors. Or if they could, they weren't saying. The "characters or letters" that Diego de Landa imagined functioned as an alphabet—they are in fact logo-syllabic glyphs: each symbol can at once represent either a word or a syllable—would not be legible again until they were decoded by the Soviet linguist Yuri Knorozov, four centuries after de Landa's reign of fire.

The grandmother failed, by the way. Hunahpu and Xbalanque heard a hawk shrieking above them and shot it in the eye with their blowguns. When it fell from the sky, they asked it what it wanted.

"I bring a message in my stomach," the hawk replied. If the twins would repair its eye, it promised, it would deliver the message.

The twins agreed. They replaced its eye with a bit of rubber, healing it instantly. "Speak," they told the hawk.

The hawk vomited up the snake.

"Speak," said the twins to the snake.

The snake vomited up the toad.

"Speak," said the twins to the toad.

The toad tried to vomit. Nothing came out. The twins kicked the toad, but still it could not vomit. They pried open its jaws and saw there, hanging from the roof of its mouth, the louse, which had not wanted to be swallowed.

"Speak," said the twins to the louse. And it delivered its message at last.

≈

There was good news too this weekend. Or what these days counts as good news. A former lieutenant general who had briefly served as the Rhino's national security advisor pleaded guilty to lying to the FBI and agreed, as part of his plea bargain, to cooperate with the special prosecutor's investigation of Russian interference in the 2016 election. The investigation, in other words, of the Rhino. He is the second and by far the most senior of the Rhino's former advisers to take such a deal. The hounds are getting closer.

This is not quite a comfort. Distraction is the Rhino's preferred tactic. Even without a crisis, it is the only trick he knows: to provoke some fresh outrage and profit from the confusion that he sows. He has surely figured out that there is no better distraction than war. Perhaps the timing was coincidental, but on Saturday, one day after the plea deal hit the papers, his current national security advisor told Fox News that the possibility of war with North Korea is "increasing every day." Military action might still be avoided, he added, but "there's not much time left."

How can we understand these metaphors? Not much time left, time is running out, as if time were a ribbon and we're getting close to the end of the spool. What black hole awaits us when the last of it unravels?

z

It turns out that jumping into a black hole is a difficult proposition. Black holes cannot be seen or directly observed, their gravitational pull being so strong that it sucks in even light. Their presence, though, and a fair amount of information about them, can be inferred from the behavior of the gases that surround them, and by their effect on the movements of other bodies, such as stars. From which we can gather that, if you were serious about this, and really jumped, as you approached the event horizon—the one-way boundary that defines the black hole, through which you can enter but never leave—you would undergo a process that physicists have dubbed *spaghettification*. (Poets do not often become physicists.) The gravity pulling on one end of your body—say, your feet—would be stronger than the gravity tugging at the other end, so you would be stretched like a nice, fresh hand-pulled noodle. If it wasn't too uncomfortable, and didn't immediately kill you, your head could watch your feet recede. Eventually, though, you would reach the infinitely dense one-dimensional point at the black hole's center, the singularity, where you, and time and space, would cease to exist in any way that we are capable of understanding.

Here's the funny bit. If I jumped into the black hole, and you stayed outside and watched me do it, you would not see me disappear. You would see me moving more and more slowly, stretching horrifically but perhaps quite comically as I went until I came to a complete stop just outside the event horizon. If I was wearing a watch and your eyes were good enough, you would see its hands spin more and more slowly until they stopped. Then you would likely see me be incinerated, watch and all, by the heat of the so-called Hawking radiation emitted by the black hole. But whatever you saw would be very different from what I experienced as I continued to float along past that horizon, stretching as I went, yawning perhaps, until I arrived at the singularity and, in this universe at least, ceased to be.

Pedro de Alvarado makes a brief appearance in a Borges story, "The Writing of the God." The plot is so simple that it hardly counts as plot: a character has a question; it is answered; the tale ends. A Maya priest called Tzinacán narrates the story from a vaulted cell deep in a stone-walled prison. The temple over which he once presided, dedicated to the K'iche' god Qaholom, was burned by Pedro de Alvarado. His cell is divided by a wall. On the other side of it is another prisoner, a jaguar that "with secret steps measures equally the time and space of its captivity." Once a day, at noon, a trapdoor opens above the cell so that the prisoners can be fed and, for a moment, enough light enters that the priest can see the jaguar.

The priest—Borges uses the word *mago*, closer to "sorcerer" or "wizard," and hence to the tradition of the Renaissance *magus* than to anything recognizably Mesoamerican—does not expect to be released from his prison. He passes the time trying to remember everything that he knew of the world. One night he recalls that on the first day of creation his god foresaw "that at the end of time there would be great misfortune and destruction," and wrote a single magic sentence capable of preventing that disaster. But "no one knows where he wrote it, nor with what characters." The priest judges that "we were, as always, at the end of time," and that it is his destiny to find the hidden script.

It could take almost any form, he knows, "a river, the empire, the configuration of the stars." It could be his own face. It occurs to him that the jaguar was among the attributes of his god and he becomes convinced that the sacred text is inscribed on the body of the jaguar with which he shares his prison. He spends years memorizing every mark on the animal's coat, but he does not know how to interpret their patterns. He begins to lose hope: what words comprehensible to the impoverished cognition of humans could begin to approximate the speech of a god?

Only after he utterly despairs does the vision come to him. In an ecstasy of mystic union, he sees an enormous spinning wheel, composed at once of water and fire. In it, interwoven like threads in a fabric of infinite complexity, he can see everything that is and was and ever will be. He sees himself, one of countless strands. He sees Pedro de Alvarado, his tormentor, another. He sees the entirety of the universe and its "intimate designs . . . infinite processes that together formed a single happiness." He understands it all, even the writing inscribed on the jaguar's flesh.

The text is composed, he says, of fourteen otherwise unremarkable words. To pronounce them aloud would be enough to give him the powers of a god, the ability to feed Alvarado to the jaguar, "to plunge the sacred knife into the hearts of the Spanish, to rebuild the pyramid, to rebuild the empire." But he doesn't do it. His vision has dissolved his belief in the centrality of his own existence. The man he had been, who craved vengeance, and meaning, doesn't matter anymore. His petty misfortunes, his people, what did any of it add up to, "if he, now, is no one"?

❧

Not everyone likes the desert. I've loved it since the first time I came out here alone—not to Joshua Tree that time, but farther north and east, to Death Valley. This was eighteen years ago. I had messed my life up in a number of ways that at the time felt irreparable. Being young and overly literal, I decided to head for the lowest spot in the hemisphere. I drove out from L.A., wrapped a scarf around my head, and walked out beneath the brutal sun over the salt flats in Badwater Basin. The earth was cracked and crusted white, the heat quivering above it. I doubt I brought much water. I didn't make it to the very lowest point. It was too hot, and walking out there, the salt crackling beneath my boots, I started laughing and couldn't stop. Mainly at myself, but at everything else too. The misery, the sense of failure that had sat on my shoulders for months just lifted off, pulverized by pure absurdity. I remember finding insects that had died there and been encrusted with white crystals of salt. Little jeweled crickets. I put a couple in my pocket and didn't feel bad about turning around. I stayed in Death Valley for a day or two longer and can't remember ever feeling so free.

Later my ex and I would rent a cabin in Joshua Tree for a month or two each year and come out to write. She had lived there before and she teased me at first: "City boy, you'll be running back before a week is up." It turned out that the solitude and silence suited me. And the clarity of the light. Sometimes I came alone for weeks at a time. I didn't even want to drive into town to buy groceries. When work forced me home to Los Angeles—and really, I loved L.A.—I raced back to the desert as soon as I was done. The moment I got off the interstate and headed up the grade and over the mountains I would roll down the windows, sniff the air like a dog, and feel the tension sliding from my spine. I grew up in the New York suburbs and my people are originally from cloudy, low-skied lands—Scotland, Ireland, Poland, Ukraine—but I had never been anywhere where I felt so immediately at home.

Some people have the opposite response. My godmother and her partner came down to visit once from San Francisco. Her partner, who was probably sixty at the time, was overcome with joy the first time we took him into the park. He turned into a little boy, scrambling over boulders, his eyes enormous, his face transformed. But my godmother felt uneasy and exposed. Everything was sharp. There was too much death around. She missed the nurturing embrace of leafy green plants, moisture, and abundance. Other friends have had the same reaction. It's not an aesthetic aversion so much as an existential allergy. They feel dread, something approaching panic. They see only emptiness and the bare cruelty of nature, though the forest and the coast are no less cruel.

I tried to explain that what I saw around me was not just death but, right next to it, sharing the same space, the urgency, brilliance, and stubbornness of life. You couldn't always see it, and never would if you didn't look, but everything was alive. Even the rocks and the dirt are alive. I don't mean that in some mystical sense. Or not only in that sense. A "cryptobiotic crust" of microorganisms—bacteria, lichens, mosses, algae, fungi—covers the desert floor, an invisible web of fibrous tentacles that allows the soil to absorb the rain and resist the wind, sheltering the roots of plants and the animals that tunnel and burrow down there, protecting everything that skitters beneath the surface of the seen. Lichens splash the rocks with brilliant greens, yellows, reds. Life thrums through this place like a current coursing through matter that is anything but inert. A barrel cactus, a brilliant, neon pink, growing alone in a crevice, anchoring itself in a few inches of sand between the rocks. Rodents that will never in their lives drink a sip of liquid water survive solely on the sparse moisture in the seeds they so nervously eat. The spring of a jackrabbit startled in a wash. The speech of ravens. And, at night, of coyotes. And of owls. In the spring, after a New York weekend's worth of rain stretched out over the entire winter, it all bursts forth in wild celebration—shrubs that seemed leafless, dry, dead for months reveal themselves in sudden and outrageous color, like drag queens at a ball. Some of them bide their time, waiting for the late-summer downpours and only then, in August or September, showing themselves in brilliant yellows, purples, blues.

Yes, death was everywhere too and more obvious here than in the fir and redwood forests of the north or the oak and chaparral that roll over the hills along the coast. There are rattlesnakes and mountain lions and the heat is surely lethal. There are creatures out of nightmares. There's a giant wasp bigger than a hummingbird that lays its eggs in the flesh of living tarantulas so that its offspring will have something to eat the moment that they hatch. There's a lizard that shoots blood from its eyes to scare off predators, a shrike that impales lizards on yucca spikes so that it can eat them at its leisure. Sometimes you see them suspended like a warning, the corpses of heretics left hanging from the ramparts.

You can't avoid the cruelty, though that's the wrong word because it's nothing so intentional. Your consciousness is not central to this drama. No one's is. That may be the real source of horror here, and of liberation. Whatever you imagine is unique about yourself, whatever you think matters, the coyotes don't care and the owls don't care and the stars most certainly don't. The desert would be fine without you. It will be. Even if in our heedlessness we wipe out half the species in it. The desert practically shouts that at you, all day and all night, that it, and life in all its resiliency and multiplicity and magic, pulsing force, will go on. Whatever we do or don't do, whether we're still here or not.

When I was an unbearable teenager—think unfiltered cigarettes, hair in my eyes, long black overcoat, a walking sulk in pegged Gap jeans—I came across a line of Samuel Beckett's that stuck with me. (This would have been in the late 1980s, when the amount of carbon dioxide in the atmosphere rose for the first time in about three million years above 350 parts per million, the level to which human beings, and most life on the planet, are adapted. I didn't know that then. Most of us didn't.) "Every word," Beckett had said in an interview, "is an unnecessary stain on silence and nothingness." This confounded me. I agreed with him, enthusiastically. I was solidly in the pro-nothingness camp. Showily, even. But Beckett had found it necessary to say those words. Ten of them in that sentence alone. Ten stains running into one, dripping all over, spoiling the virgin purity of the void. It angered me, as if Beckett had betrayed a vow. But I held him in sufficient awe that I figured he must have had his reasons, and this was reassuring. It let a little light in. Beckett had validated the urge to speak, to write. He couldn't stay silent about silence. Words are self-propelling. They boil up. The void coughs them out, like hair balls. Or the pellets regurgitated by owls: bones and hair, whole undigested limbs sometimes, teeth, feathers, claws, everything the gizzard can't digest.

That's the thing about silence I would tell my teenage self if I could: it's very loud. And nothingness is teeming. I say this here because Borges made some version of the same move in "The Writing of the God." The priest's vision convinces him of his own insignificance, of the insignificance of any man, being, nation. That's why he doesn't pronounce the holy formula, he admits in the story's final sentence, "that's why I let the days forget me, lying in the dark." But Borges made him say this. He wrote the story, published it. He created the priest as a character, in prose, so that he might attest, in lasting print, to the futility of words.

❧

I'll admit it's not my favorite Borges story. It is perhaps not a very good one at all. Borges's best tales fold in on themselves vertiginously, losing the reader in the same paradox that dissolves the narrative foundations of the story itself. This one simply dissolves. The ostensible plot, the historic Maya frame, reads like a stage set hastily constructed in anticipation of the main act, a mystical revelation to which the setting bears little relation. Borges appears to have borrowed the name Tzinacán, which means *bat*, from a Maya chieftain (not a priest) who is mentioned just once in Bernal Díaz de Castillo's *True History of New Spain*. The name of the god who writes the hidden sentence, Qaholom, comes up, by my count, five times in the entire *Popol Vuh*. Any other exotically indigenous name would have served him just as well.

Borges was impatient to get to that wheel. He may have seen some echo of it in the calendar stones carved by the Aztecs, which at least are circular and have to do with time, but the image is more likely borrowed from his readings of Buddhist and Kabbalistic texts, and of the Renaissance mages and mystics who borrowed from the latter. Like any good Argentine bourgeois, Borges looked across the oceans for profundity. Time existed elsewhere. In his writings he displayed little cognizance that the hemisphere on which he lived had a history of its own that preceded the arrival of Europeans. It may be crude to suggest this, but it is not hard to read in "The Writings of the God" a strangely labored justification for silence in the face of the genocide in which Pedro de Alvarado took part, a silence that would later echo through Borges's quiet support for the military dictatorship that dominated Argentina in the 1970s and '80s. How else to understand, in the hands of an otherwise so meticulous writer, the flimsiness of its construction? Why was it so important that this vision of the insignificance of human striving be voiced by history's vanquished?

2

I went out for a run earlier, heading west along the boundary line of the park and toward the setting sun. I was maybe a half mile from the house when a coyote skipped across the road in front of me. I must have surprised him. If I hadn't I'm sure I wouldn't have seen him at all. He glanced at me sideways and jogged on without slowing, rendering himself invisible among the creosote and senna.

I was probably more surprised than he was. I hear coyotes every night, but I sometimes go weeks without seeing one. They're around, of course. They're just good at not being seen. Usually they start singing at dusk or a little before, a single yipping voice, then others join in, echoing one another and coming together in a chorus that rises to a frenzied, ecstatic peak and just as suddenly dissolves. If it's still light and they sound close enough—sometimes they must be yards away—I go outside and look for them. I never see them. They blend into the desert too perfectly. After midnight their calls break through the darkness and I often wake and wonder what it is they're hollering about: a successful hunt or a failed one, a jealous squabble, or something less dramatic. Maybe they're like us and just need to hear themselves, and one another, to find some way to fill the hollows of the night.

I've been seeing them a lot lately. I saw one in my headlights just down the street last week, and the other day in the afternoon the neighbor's dogs all began barking at once. When I looked out the window I saw a coyote trotting down the middle of the road. I find myself doing it again, putting myself at the center of this story: I flatter myself that they know I'm leaving soon, that they've been listening in on my phone calls or pressing their ears to the screens when I talk to L. on the phone, that they just *know*, and this is their way of saying goodbye. But there is no story here, or, what amounts to the same thing, there are infinite stories, with infinite centers, and coyotes surely have other things to think about.

Perhaps the least convincing thing about "The Writings of the God" is that the priest's vision leads him into a dull quietism, as if he had seen that spinning wheel from the heights of a mountaintop and decided that the cosmos was populated only with the tiniest and most insignificant of beings, paramecia and plankton, and that his own death and the rise or fall of his or any people was of no more consequence than the fading of a spot of lichens from a stone. He wasn't wrong about that, but he only looked through one lens of the telescope. The other side—the one that makes things bigger—is even more interesting, if more painful to take in. If you can blink through the tears and focus, you'll see that secret words are written everywhere, on every hair and every cell and every star.

Borges was wrong. The gods don't want us to lie down. They don't want to watch us vacillating, blinking, stuttering. They like to see us dance and fight. They like to watch us act with grace and conviction. They want us to read what they have written. They want us to pronounce the secret words aloud.

≋

Yesterday I read a warning that the Santa Ana winds would be blowing hard through L.A. I haven't been back there for a few weeks, but firestorms have been raging for days in Ventura, north of the city, burning more than one hundred thousand acres, from the mountains to the sea. Wildfires are normal in Southern California. This is not. The rains are late. Usually they arrive in October, bringing the fire season to an end. It's December now and it hasn't rained since May. But there is no *usually* anymore. Except for the odd wet year in which the pendulum swings to the other extreme, the rains have been coming late. The fire season has grown about a month and a half longer than it used to be. I'm not that old—I turned forty-five this fall—but I lived in Los Angeles for nearly twenty years, long enough to see its climate shift.

Now this is it. The disasters they warned us about are here. The future has already happened. Last year was that odd year, the wettest on record for the entire state. The spring was glorious—I've never seen so many flowers here—but it all dried out and the hills up and down the coast are thick now with kindling. This after several years of drought. The driest year on record came just two years earlier, leaving behind millions of dead trees. All of which adds up to perfect conditions for uncontrollable firestorms. This, according to the climate scientists, is the way it will go. Decreased rainfall in California is tied to loss of Arctic sea ice. The dry years will be drier and the wet years, when they come, will be wetter. Everywhere it's hotter.

By this morning the winds had done their work. These ones are blowing from east to west so there's no sign of the smoke from here, but I watched a video on Twitter that someone shot while driving to work in the predawn dark. The four even lanes of the 405 freeway were familiar enough. So was the sign indicating that the Getty Center exit was a half mile away, Sunset Boulevard 2½ miles, Wilshire 3¾. The car was driving south into the wealthiest part of the city's west side. Through the smoke you could see the flames covering the hillsides to the east in orange and a blinding yellow-white. It looked like Mordor.

*

Perhaps the most elegant and nightmarish of all of Borges's stories is "The Library of Babel," in which he imagines the universe as an infinite, hive-like library of largely incomprehensible books that together contain everything that can possibly be expressed in all possible languages, known and unknown, as well as a great deal of actual babble. I was thinking about the story recently when I came across a post about black holes on Stephen Hawking's website. This is the bit that got my attention: "One can't tell from the outside what is inside a black hole, apart from its mass and rotation. This means that a black hole contains a lot of information that is hidden from the outside world. But there's a limit to the amount of information one can pack into a region of space . . . If there's too much information in a region of space, it will collapse into a black hole . . . It is like piling more and more books into a library. Eventually, the shelves will give way, and the library will collapse."

Borges's narrator was, in his way, more optimistic. "I suspect that the human species . . . is about to be extinguished," he wrote, "and that the Library will endure."

*

Black holes, by the by, are just collapsed stars. They are sites of haunting, the forces that worlds continue to exert after they cease being worlds. The tug of the past so strong and furious that it breaks down time itself. And they shape everything. Most if not all galaxies swirl around black holes. Ours does. At the center is a void, impossibly dense, a nothing that is teeming with being.

❧

When I was nineteen I narrowly escaped from what was, until this fall, the deadliest wildfire in California history. It was outdone this year, in October, by fires that raged through five counties in Northern California, killing forty-four people. It was October then too. Two friends and I had driven from the East Coast to the West until our transmission blew as we crossed into California from Oregon. We spent a few days in a motel in Crescent City waiting on a rebuild. Crescent City was foggy and dull, we were running out of money, and the mechanic was moving slowly, so we hopped a bus to Oakland, where we had a place to stay. A beautiful place: my godmother's home in the hills overlooking the San Francisco Bay.

We lazed over coffee that first morning, enjoying the view, puzzled by a strange orange tint to the sunlight. The air smelled of smoke. My godmother called the fire department. They told her not to worry: they would let us know if the fire got too close. Just in case, she tossed me the keys and asked me to pull the car from the garage. By the time I had backed it into the street I could see flames crawling down the hillside. By the time we drove off, less than ten minutes later, smoke blocked the road in both directions. Everything was burning.

The memories feel like a dream: we ditched the car and ran down the slope, flames spreading through the underbrush and licking the trees all around us. We made it to a clear stretch of road and caught a ride sitting on the trunk of someone else's car. He wanted to drive faster so he kicked us off halfway down. We caught another ride and made it somehow to Telegraph Avenue in downtown Berkeley. Everyone was standing in the street, gawking up at the burning hills, the weird orange ball of the sun staring back down through the smoke. Someone tried to sell me acid. The fire killed twenty-five people that day.

I went back with my godmother when the city at last let residents through the roadblocks to inspect what was left of their homes. Only her chimney was left. The car, reduced to skeletal essentials, was hundreds of feet from the spot where we had abandoned it. The gas tank must have blown. The neighbor's Jaguar was gone too, in its place a few small puddles of chrome that had dripped off the bumper and grille then pooled and congealed on the ground. We found the living room: the books that had lined the walls from floor to ceiling had been transformed into an undulant white sea of ash. It was quiet up there—there were no birds anymore, and not a leaf to rustle—and it's possible that I had never seen anything so beautiful. I could make out individual bindings and the deckled edges of pages that had once borne words as clear as these ones. They collapsed as soon as I touched them into a fine, slippery powder.

3.

Driving home the other night I saw a shooting star, huge and green and straight ahead of me, streaking so low across the sky that it seemed to hang there for a moment before it faded. I thought for a second that it must be a stray rocket from the marine base just off the highway in Twentynine Palms, and perhaps it was. Hegel, the German philosopher who wrote a great deal about history and the nature of change over time, also wrote about strange things that fly through the early evening sky, and about owls. It's one of his catchier lines, and certainly his most famous, from the preface to *Elements of the Philosophy of Right*: "The owl of Minerva begins its flight only with the falling of dusk." Minerva being the Roman equivalent to Athena, the Greek goddess of wisdom, weaving, and war, who at times took the form of an owl. What Hegel meant is that wisdom comes too late. Always. ("It is only when actuality has reached maturity that the ideal appears opposite the real and reconstructs this real world, which it has grasped in its substance, in the shape of an intellectual realm.") At any given moment the froth and swirl of events can only blind us and confuse us. There is no way to get above them from within the confines of the present. Only after the fact, when night is already falling, are we able to look back and understand.

We might now look back, for instance, to 1820, when Hegel first published those words, when coal-powered steam engines were beginning to replace water wheels in the busy textile mills of northern England. (Germany and the United States would not turn to coal until after the transition in Great Britain was complete.) But for that owl and its nocturnal habits, we might, with all the smugness of hindsight, insist that the black smoke that spilled from their chimneys and filled their lungs should have given early industrialists a potent clue that this would not end well. And we might remind ourselves that the owl will fly again tonight, and again at dusk tomorrow, and that none of this has ended yet.

Hegel, in any case, looked back and saw something like God, which he called Spirit. Human history was for him a rational process, and also a divine one. Subjected to the vigilance of philosophy, the logic propelling it would reveal itself, but only after the fact. History was the very thought of God as it developed over time. It formed a single epic narrative, the story of the growing self-consciousness of Spirit, of God coming to know himself, through us, in time. The trajectory was clear: from slavery to freedom. (Not incidentally, this could also be expressed geographically: "World history travels from east to west; for Europe is the absolute end of history, just as Asia is the beginning.") The latter arrived, for Hegel, in the perfection of the modern state, which, he wrote, "is the realization of Freedom, of the absolute, final purpose, and exists for its own sake . . . The state is the divine Idea as it exists on earth."

One hundred and twenty years later, Walter Benjamin, a different sort of German philosopher, saw things differently. Months before he ended his life in a hotel room on the French-Spanish border, despairing of an escape from the Nazis, he wrote twenty paragraph-length fragments on "The Philosophy of History." In the ninth, the most famous of them, Benjamin described "the angel of history" being propelled blindly into the future, still facing the past: "Where we perceive a chain of events, he sees one single catastrophe which keeps piling wreckage and hurls it in front of his feet." The angel is pushed onward, Benjamin wrote, by a terrible storm. "This storm is what we call progress."

≈

Yesterday the Rhino—the perfection of the modern state—recognized Jerusalem as the capital of Israel and promised to move the U.S. embassy there from Tel Aviv. The announcement was condemned by pretty much every government in the world except Israel's. I spent the morning trying to figure it out. Surely there was some rationale for his recklessness, if not a strategy then at least some sleight of hand. As far as I can tell there wasn't. To please his most rabidly right-wing and pro-Israel donor (the Las Vegas casino magnate Sheldon Adelson) the Rhino had made a promise during the campaign. He was convinced that keeping it would please not only Adelson but the evangelical Christians who form a large part of his base. He is said to have refused to explain his decision to the Palestinian president, telling him only that "he had to do it." Unnamed aides told *The Washington Post* that the Rhino "did not seem to have a full understanding of the issue." So we race into the abyss.

I record this not as evidence that the Rhino is particularly stupid, shortsighted, addled, deluded, demented, arrogant, venal, and vain, though he is all of those things. And night is dark, the sun hot, and bright. The Rhino's election and the recklessness with which he rules are not potential causes of global chaos but symptoms of a breakdown that is already under way. A healthy, confident nation would not have elected such a man. This one is sick to the bone, stumbling everywhere it steps, knocking things over, making a mess. Empires do not go down quietly. Usually they take the whole world with them for a while. The last great hegemonic handover, from Great Britain to the United States, followed two world wars and the loss of many millions of lives, most of them neither British nor American.

The Rhino, with his vituperative, uncomprehending eyes, his puckered lips and painted orange scowl, is the face of this collapse. He's what we look like now. Every buried crime and contradiction on which the American polity was built sprawls in the open over the sidewalks and the streets and the endless crawl of cable news. The Klan is out of hiding. The dumb ones wear swastikas. The rest, in suits and ties, strut the soft-carpeted corridors of power. The rich are stealing everything. They don't bother to hide the graft or to disguise the contempt in which they hold us. All the sexual horrors are spilling out, hungry priapic wraiths with sticky palms and iron grips haunting every workplace. This is what we look like. Nothing remains concealed. The past is returning. The unconscious won't stay un. And into this cauldron the hurricanes and fires blow, one after another.

Or think of it this way. Hundreds of millions of years ago, our distant cousins—various phytoplankton and zooplankton, cycads and ferns—lived lives as full of passion and drama as any, and then went ahead and died. Buried in mud or water and deprived of oxygen, they were compressed over the centuries by layer after layer of sediment and stone. Slowly, pressure and heat transformed them into a black and viscous goo, into gas that stinks of flatulence, and into strange, hard, oily lumps. Cut to the early nineteenth century, when British industrialists found a use for these otherwise unpleasant substances, the transformed bodies of the earth's early dead. They burned them, and made things move, and turned that motion into money, which could be turned into more money to mine more ancient fuels from the earth and make more things move and make more money. The carbon that had for millennia slept beneath the planet's crust in vast and oozy subterranean cemeteries was suddenly spat into the air through smokestacks, chimneys, exhaust pipes. It stayed up there and commenced absorbing more and more of the radiant heat of the sun, causing the earth to precipitously warm, the ice at its poles to melt, its oceans to rise, their currents to shift. You are no doubt by now familiar with this process. What is it, really, though, but a haunting—the ancient dead disturbed from slumber, punishing us for our greed and blindness, our restless lack of reverence? What is it but the past come back, and time unhinged, collapsing?

≈

Walter Benjamin attributed the failure of the social democratic politicians of his day to reckon with the threat of fascism to their "stubborn faith" in progress. If mankind was destined to advance, how could the fascists, with their crude and backward-looking talk of blood and soil, be taken seriously? But they were serious, and so are their descendants today.

The problem for Benjamin was not simply that faith in progress was mistaken. It was that the entire idea relied on a concept of time—a time that was "homogeneous" and "empty"—that was itself illusory, and dangerous. The opportunities and hazards of the present, Benjamin argued, could not be understood unless time itself was reconceived.

If this was true then, how much more so is it now, when fascism is not the only peril that we face?

≈

Last night, by a hair, and against the wishes of nearly 70 percent of the state's white voters, Alabama failed to elect to the U.S. Senate a man who spoke nostalgically of slavery and who was banned from his local mall for preying on teenage girls. Time is not moving smoothly forward. It's circling back, getting knotted up in oblong loops, stopping, stuttering, plunging on.

In the summer of 2014 I was living in Ramallah. It was a very bad year. War didn't break out until early July, but for most of June Israeli troops had been flooding the West Bank. The days were long, the nights even longer. I don't remember sleeping much, only lying in bed, listening to the dogs bark, waiting for the call to prayer to announce the arrival of the dawn. The clashes at the checkpoints started in the afternoon and stretched late into the night: boys and young men throwing stones at soldiers who fired back with tear gas canisters, rubber-coated bullets, live ammunition. No one flinched at the blasts. The young men took breaks from throwing stones to direct traffic and smoke cigarettes, trying to keep the city flowing. Later, when everyone but the kids standing watch outside the refugee camps had gone to sleep, the soldiers came into the city to raid houses and make arrests. Shots and explosions shattered the night. Each morning's news was worse than the last's. Then the war started. Too much was happening, all of it bad.

Time seemed to have changed its shape. The clocks behaved as they always had, ticking away, counting off the hours. They seemed to mock us. Time no longer proceeded evenly and sequentially, but according to a strange logic of dread. It curved and bent, revealing pockets inside itself, pockets and holes in which it was easy to get lost. Sometimes time rushed forward, then something happened—usually death—and it stopped, melted, and recovered. It lurched off, racing once more, zigging and zagging before dissolving again and somehow, from nothing, reconstituting itself and limping on.

I had felt this before in other countries on the verge of collapse. I've felt it since, not quite so acutely but nearly constantly, in the year since the Rhino's election. I don't know what to call it. The Time of Crisis, Vertigo Time, the Time of Collapse, Black Hole Time. The days and hours lose their shape, their uniformity, the confidence with which they once marched forth. Time appears to fall apart.

*

For the Romans too, and the Greeks before them, owls were messengers. Better put, they were a glimpse of the goddess herself. Sometimes symbols are the very thing. Athena's human form was no less a mask than her owlish one. She was the patroness of Athens, so Athenians, proud inventors of democracy, stamped owls on their coins and branded them on the faces of slaves captured in battle. (That's a lot of owls: per the classicist Moses Finley, slaves accounted for as much as one-third of the population of Classical-era Athens.) Throughout the Mediterranean, the goddess appears on pottery and in sculpture standing beside an owl, or holding one, or with an owl on her head. Wisdom takes some funny shapes. Sometimes she had an owl's wings and talons growing from an otherwise human female form, or an owl's body and a human head, helmeted and ready for war.

Over the centuries and throughout classical literature, owls meant one thing—trouble—unless you were lucky enough to be from Athens. Plutarch wrote of an owl alighting on the mast of an Athenian ship before the battle at Salamis, lending the Greeks the courage to defeat the Persians. The tyrant Agathocles is said to have released owls over the ranks of his army to convince his soldiers that the goddess was with them. In Aristophanes's *The Wasps*, an owl flies over the Persian troops just before the fighting commences, a sign that the Greeks would triumph. So complete was the association that the bird became a proverb: according to the British classicist Arthur Bernard Cook, to observe "there goes an owl" meant that victory was close. But, Cook cautioned in a footnote, "The bird which portended victory to friends naturally portended defeat to foes. Consequently the owl also had a sinister significance."

The owl is always ambiguous. Archaeologists have dug up pendants in southern Italy showing Athena with an owl's wings and human hands, which she uses to spin wool into yarn. Weaving, wisdom, war: How can one deity oversee such disparate charges? In "The Writing of the God," Borges also described that whirling wheel of time as a fabric embroidered with impossible complexity. (The words *text* and *textile* are both from the Latin *texere*, to weave: writing is, perhaps first of all, woven, a fabric of overlapping threads.) His imprisoned priest glimpsed the entire weave at once without any of the comforting lies of narrative, without cutting it down to a single and seductive swathe that, once chosen, negates all other possibilities and obscures the remainder of the cloth from which it's spun. But it's all still there, even when we fail to see it. Pull any thread and you'll tug another that you didn't mean to move. You'll find entire worlds. In some of them gods could be birds and birds gods. Homer depicts Athena as a pigeon, a swallow, a hawk. In the *Iliad*, she and Apollo appear as vultures perched high in an oak tree to watch the Greeks and Trojans battle. They like to watch us fight.

The archaeologist Marija Gimbutas saw in Athena an incarnation of a much older divinity, which she called the Snake and Bird Goddess of Old Europe. Gimbutas, who was born in Lithuania and had to hide during the war from succeeding military occupations by the Russians and the Germans and the Russians again, had a brilliant but fairly conventional career until the early 1970s, when she began to write about goddesses. In 1974 she published *The Goddesses and Gods of Old Europe* and first laid out the hypothesis that she would continue to elaborate until her death twenty years later. It began by proposing that there was such a thing as Old Europe, a distinct and sophisticated Neolithic culture that stretched from what is now Ukraine and the Czech Republic to the northern shores of the Mediterranean, one that did not owe its achievements to the more storied civilizations of Mesopotamia and the Levant. The societies of Old Europe, Gimbutas was certain, were matriarchal, egalitarian, and pacifistic, and centered on the worship of a nurturing goddess. All that was destroyed, she argued, by invaders from the east: fierce, equestrian Indo-European nomads who replaced the earth-oriented and goddess-centered pantheon with cruel male gods of sky and storms, and who brought with them patriarchy, hierarchy, exploitation, and war. What little of Old European culture was able to survive would be forced into subterranean channels.

All this was widely dismissed by archaeologists at the time, many of them making the sort of complaints—that Gimbutas's conclusions were irrational, sentimental, and insufficiently rooted in empirical evidence—that men tend to make when women say something they don't like. ("Most of us tend to say, oh my God, here goes Marija again," Bernard Wailes of the University of Pennsylvania told *The New York Times*.) But feminist archaeologists would also find much to criticize in Gimbutas's ideas, which were most wholeheartedly embraced outside of the academy, by New Age feminists hungry for alternatives to the patriarchal and militaristic society in which they lived. In which we still live.

Looking back, it's hard not to find something sinister in the narrative of a lost utopia that she imagined—a genteel and gentle Europe assaulted by brutish outsiders invading from the East. If only structurally, it too closely echoes the primitivist fantasies of the latest generation of purity-obsessed ethnonationalists. And Gimbutas surely gave too little credit to the goddesses, and to actual women, stripping them of all but the most stereotypically maternal aspects of human personality. Athena would not have easily forgiven Gimbutas for suggesting that she had been transformed by corrupting foreign influences from a nursing, protective mother god into a goddess of war, and that she only became capable of ferocity and wrath after being "Indo-Europeanized and Orientalized during the course of two millennia of Indo-European and Oriental influence in Greece." It's a bit like Botox: Athena's youthful beauty is restored here, but she's no longer able to scowl.

Still, I can't help but find myself circling back to Gimbutas. However loopy the details, in broad outline much of what she wrote seems right. They may not have organized themselves into model feminist communes, but for many, many centuries and until quite recently, humans all over the planet did worship goddesses, and then they stopped. Most of them anyway. Implicitly if not explicitly, under both monotheist and rationalist conceptions, the cosmos is gendered male. This shift seems worth thinking hard about: what it means, what was lost, what might be worth recovering. Someone else can take that up.

What keeps drawing me to Gimbutas is her combination of the darkest apocalypticism and an optimism that, though it is only two decades distant, feels at once difficult to salvage and, in some basic sense, essential to our survival. In her telling, time has a different shape. It's not a vector pointing upward that is suddenly, cataclysmically collapsing. The disaster already happened. It came in hordes from the East, on horseback, carrying cruel gods and weapons of bronze. Nothing was left standing. Doomsday came and went. It happened so long ago that we've forgotten it, repressed it, hidden it from our collective memory. This notion—perhaps more than any imaginative overreaching and selective marshaling of archaeological evidence—may have been what put Gimbutas on the outside of the academic mainstream: she is saying that we got it all wrong. Civilization as we know it is not an achievement, but a tragic defeat. Most of what we recognize as history was founded on a catastrophe that has only been compounded with the accretion of the years. But this also means we are not damned to this, that there are other ways to live, that we have far less to lose than we thought we did, and a great deal still to learn.

"We've reached the end of the world," Gimbutas said in a 1990 lecture. "We're starting to create another. I expect we shall become a healthier society. We shall worship the earth—well, not in the same way, nothing returns from the past. We cannot repeat the whole thing from the beginning, we can only transform ourselves and use our knowledge about the past and apply it for creating the future. This is my feeling."

Long after the era that obsessed Marija Gimbutas but more than a millennium and a half before James Watt patented his steam engine, propelling the mills of Great Britain into the feverish consumption of coal and the planet into the current era of cataclysmic climate change, the island's Roman occupiers were already, on a far smaller scale, digging that miraculous, slow-burning black stone from the coal beds of England and Wales. They used it in smithies, to forge the weapons and armor that allowed their empire to advance; they used it to keep warm through the wet English winters; and they used it to fuel the eternal flame that they kept burning in Bath, in a temple erected there to Minerva, the wisest of the gods.

Athena was not always admirable, but then gods are notoriously proud. Arachne, a common girl from Lydia, now somewhere in western Turkey, was a weaver like Athena, and grew famous for her talent with the loom. In Ovid's version of the story, Arachne was also proud. She did the unthinkable and challenged Athena, whom Ovid calls Minerva. The two competed, the goddess and the girl. Athena wove an image of the gods arrayed in all their majesty, and embroidered in, as a reminder, the fates of various mortals foolish enough to challenge them, transformed into birds or trees or icy mountaintops. It communicated all that power ever wishes to, seam-lessly, at once propaganda and threat. (Ask Hunahpu and Xbalan-que: the messages of the powerful are always invitations to submit.) Arachne, defiant, wove the gods as she saw them, as deceivers, rap-ists, thieves. The beauty of her tapestry exceeded even Athena's, and it achieved something the goddess's could not: truth. Athena tore it from the loom and thrashed her. Arachne, despairing, would not consent to be humiliated. She would not kneel. She hanged herself instead. The goddess, unyielding, turned her into a spider, that she might forever weave, and hang, in warning.

≈

L. is home. She flew in yesterday. In June she got a job overseas. Since then we've been apart far more than we would like to be. Fri-day afternoon I drove into L.A. to pick her up. The sky was yellower than usual. The fires are still burning, and spreading. I spent the night at S. and D.'s. The timing was good. When I got there they were making tamales for the holidays. Their daughters hadn't shown up yet, but D.'s mother was over, standing by the stove, monitoring one pot of chicken and another of pork. I helped for as long as they let me, taking a position on the assembly line, spreading masa onto corn husks, smearing the corn paste with a dollop of meat and sauce or a sliver of cheese and a couple of *rajas*, tying them shut. When I left for the airport, D. sent me off with a dozen.

L.'s flight got in late and it was nearly midnight by the time we made it out of the airport. I drove straight to the desert, L. dozing beside me, KDAY on the radio keeping me awake. When we had left the interstate I opened the windows and let the cold air fill the car. It smelled of creosote. L. woke up grinning. There was no moon. The desert was dark and the stars were bright. L. stared up through the windshield, pointing and calling out the names of the constellations as she spotted them, like old friends she hadn't seen for years. I tried to lean over the steering wheel to see them too but the highway was twisting as it climbed up through the hills and she punched me in the leg so I kept my eyes on the road until at last we turned onto our street and passed the barrel cactus at the end of the unpaved driveway. I remembered to swerve to avoid the anthill and we got out of the car and stood in the cold, necks craned back, shivering a little, our eyes adjusting, holding each other when we got too dizzy to stand.

❧

My grandfather, in his later years, developed an interest in astronomy. He bought a roll-up screen and a slide projector and whole carousels of images of the planets and the galaxies. He was a big, lumbering man. I remember him, well into his evening gin, shouting at my sister and me to shut up and sit still during a mandatory after-dinner slide show. Was he the one who bought me that little blue paperback astronomy primer that I have carried with me every time I've moved for thirty years? I still have it in a box somewhere. I never read it. He gave me a poster too, of a supernova remnant in the constellation Vela, wisps of brilliant pink and blue folding over and into one another, giving the blackness space, volume, and depth. As a kid it looked to me like a man, broad-shouldered, tall, and slightly stooped. Like him. It's up on the wall of the office I still rent in L.A., faded, its corners scarred by the thumbtacks that have held it to other walls in other cities where I've lived.

I did wonder a lot, as a teen, when he was at his worst, how it was that he had let his world narrow so precipitously while at the same time directing his gaze to the outer expanses of the universe. As if there were a balance that had to be struck. After the last and worst of several drunken car crashes, his decline was quick. Within a couple of years he could not perform the most basic cognitive tasks. He couldn't recognize his own children, or at least not the only one who still spoke to him, and couldn't access the words he needed to ask for the simplest things. My mother lost him once at Denny's and found him in a corner, pissing in a potted plant. Dementia eventually cured his alcoholism. He forgot that he drank, forgot that he had smoked two packs of Camels every day for the previous half century. When paranoia and confusion did not propel him into rages, he was gentler and more affectionate than I remember him ever having been before.

He had once been a man who valued intelligence above all other qualities. The greatest, grudging compliment he had known how to pay was to describe someone as "pretty smart." Toward the end he took to sketching out the solar system—Mercury, Venus, Earth, Mars, Jupiter, Saturn, Neptune, Uranus, and Pluto, which had not yet been demoted—labeled, in order, and to approximate scale, with a spot on Jupiter and a ring around Saturn. At one of the homes in which he spent his final years they must have had an arts and crafts room because he gave me a stiff sheet of cardboard painted in tempera with the planets in their orbits against a light blue background. It's possible that I still have it somewhere. But even when he couldn't name them anymore, he kept drawing them in their concentric ellipses, in ballpoint or pencil on napkins or envelopes or whatever scrap of paper he could find, Saturn always with its ring.

☙

For most people on the planet, for thousands of years, owls have meant only one thing. In *The Faerie Queene*, Edmund Spenser called the owl—though he called it an *owle*—"deaths dreadfull messengere." Two centuries earlier, Chaucer had written, "The owle al nyght aboute the balkès wonde, / That prophete ys of woo and of myschaunce." *Balkès* being a Middle English word for the beams that stretched from one wall of a house to another. Six centuries after that, the association between owls and ill fortune made it to suburban Long Island. I don't remember ever hearing it from a reliable adult, or an unreliable one, but I grew up with the superstition that if an owl roosts on a house, someone within will die. Not that I ever saw any owls on Long Island.

Only once have I seen an owl roost on a building. It was in 2003, at an artists' residency in the hills across the bay from San Francisco. One afternoon I spotted an owl perched on a dormer above a second-floor bedroom in the house where I was staying. I saw it there again a few days later. I remember feeling mildly alarmed for a moment, but that I know of, nothing out of the ordinary has befallen anyone who was living in the house at the time. Or nothing so ordinary as death. A little more than a year later, though, I got a phone call about a woman who had been staying in precisely that second-floor bedroom a week or two before the owl appeared. Perhaps it had been there then as well. She was in her thirties when I met her, healthy and strong. She was the girlfriend of a close friend of mine. They were planning to get married. He called me, distraught, from overseas. She had come down with what seemed to be a simple flu. Quite precipitously, her fever grew worse. She died before he could get her to a hospital.

≈

The daily spiral. The Rhino's ambassador to the United Nations held a press conference at a military base in Washington, D.C. She stood in front of a cylinder of rusting metal. It was, she claimed, an Iranian missile that had been fired by Yemeni Houthi fighters at an airport in the Saudi capital. "When you look at this missile," she said, "this is terrifying, this is absolutely terrifying. Just imagine if this missile had been launched at Dulles Airport or JFK, or the airports in Paris, London or Berlin."

Her performance was a shabby reenactment of Colin Powell's

2003 speech to the UN about Iraqi weapons of mass destruction, as if we had all forgotten that one, and that the Bush administration had lied us into a catastrophic war. This time it read more as satire than sequel. "You're going to see a rapid flow of other things," she promised. It would be funny if it weren't so frightening.

Another new study, this one predicting that the oceans will rise 1.5 meters by the end of the century, submerging "land currently home to 153 million people." The same researchers published another report three years ago. Their worst-case estimate that time was the same as their midrange estimate this time. Yesterday's panicked fears are today's sober expectations.

Marija Gimbutas was not the first one to spot the shadow of an older goddess hanging over Athena. In the third volume of his *Zeus: A Study in Ancient Religion*, the British classicist Arthur Bernard Cook proposed that Athena may in fact have been "a pre-Greek mountain mother of the Anatolian kind," by which he meant something like the goddess Cybele, whose cult lasted into Roman times. Cook found what he believed to be "a curious confirmation" of this claim in a four-thousand-year-old Babylonian clay tablet, of which he had seen a photograph in the newspaper. This was in 1936. The world was about to explode, but it hadn't yet, and Cook, who taught at Cambridge, didn't have to travel far to see the original. A century of imperial looting had its advantages: The tablet was in London, in the private collection of a Mr. Sydney Burney. Nineteen and one-half inches high and "in a state of almost complete preservation," it depicted a nude, winged goddess flanked by two owls, an apparent forebear of Athena. She had talons for feet and stood atop two crouching lions. The expression in her eyes was knowing, and defiant.

Cook was puzzled. The nudity of the goddess, he conjectured, suggested Aphrodite or perhaps Ishtar, also known as Inanna, the Mesopotamian goddess of eros and war. The lions, though, hinted at Cybele. Cook had received a letter from a colleague, the Assyriologist Sydney Smith, who speculated that the goddess on the tablet was more likely one of the nocturnal spirits associated with storms and wind that the Babylonians called Lilitu and the Hebrews called Lilith. With that latter name she would have been known to any educated Englishman of the era: Lilith had made an appearance in Goethe's *Faust* as Adam's first and disobedient wife, a beautiful deceiver, and was later taken up by Victor Hugo and by Browning and Rossetti. She was the bad girl of nineteenth-century painting and verse, and a very bad girl indeed: longhaired and lovely but a seducer of men, murderer of infants, sower of miscarriage, death, and disease. She would make a brief appearance in Joyce's *Ulysses* too, as "patron of abortions."

By then Lilith would have come to stand in for every conceivable evil that men could think to pin on female sexuality. Dark, irrational, and corrupt, she was the fetid, tangled underside of bright, right-angled, Apollonian modernity. The Assyriologist Smith was troubled: Could such a creature truly be an antecedent of Athena, whom the Greeks, inventors of philosophy, logic, and rationality, had venerated for her wisdom and virtue? "To establish a firm connection between Athene and the goddess of the plaque," Smith worried, "will it not be necessary to show that the goddess was not originally, as later, representative of Law, Liberty, and Reason, but a local demon who fell upon the transgressor (witting or unwitting)?"

It's funny, isn't it, how much we cannot see? By 1936, after one world war and on the cusp of another, and after centuries of imperial slaughter in the colonies, it should not have been difficult to imagine that daylight might be bound to night, that reason, law, and liberty were also forces of great and chthonic violence. Walter Benjamin saw this with a clarity that must have been excruciating when, three and a half years later, just before fleeing Vichy France and taking his own life in desperation, he wrote those twenty short "Theses on the Philosophy of History." It would be his last completed work. "There is no document of civilization," he wrote, "which is not at the same time a document of barbarism." That line is on his tombstone now.

But delusions are often dearly held, and nowhere more than in empires that have not yet fully crumbled. Were not the British, as the Greeks had been—and as Americans have been—the globe's sole legitimate possessor and exporter of law, liberty, and reason? Could such treasures be inherited from a mere local demon?

❧

Yesterday and the day before it was quiet, but about an hour ago the crinkled yellow leaves from the chinaberry tree outside the house began chasing each other across the ground in angry little circles. Now the wind is screaming and all the creosote bushes are thrashing about, rioting. Sometimes at night it sounds almost human, keening in the dark, drowning out the coyotes and every other noise. It can be unsettling, but hearing it and beginning to understand it—how the wind pushes the sand and carves the rocks and shapes the land over long millennia—has been one of the joys of living here, some awareness of those slow processes, the intimacy of geologic time.

It's nearly Christmas and it still hasn't rained. The fires are still burning outside L.A. The big one up in Ventura has spread to 272,000 acres, an area larger than that of Berlin, Bangkok, Madrid, or Seoul. After more than two weeks it's only 60 percent contained. The winds are picking up there too, the same hot, dry winds that blow through the deserts here.

❧

Perhaps Lilith can help explain how we got here. Or at least give us a better idea of where *here* is. The first mention of her name appears in the ancient Babylonian *Epic of Gilgamesh*, the oldest human story preserved in writing that we are able to read. Specifically in the Akkadian language preserved in cuneiform script, pressed with a wedge-tipped stylus into clay tablets that have been dated to the eighteenth century B.C. Lilith's appearance in the epic is brief. She has made her home, we are told, in the trunk of a Huluppu tree. She is not welcome there. (*Huluppu* is usually translated as *willow*, though it is not, presumably, *Chilopsis linearis*, the fragrant variety that grows in the washes of the Mojave Desert.) The goddess Inanna, also known as Ishtar, wants the tree for its wood, to build herself a throne and a bed. The story ends badly for the tree, and for Lilith. The gallant Gilgamesh cuts down the Huluppu. Inanna gets her furniture, and Lilith flees into the desert.

She shows up much later in the introduction to the first volume, published in London in 1903, of Reginald Campbell Thompson's *The Devils and Evil Spirits of Babylonia: Being Babylonian and Assyrian Incantations Against the Demons, Ghouls, Vampires, Hobgoblins, Ghosts, and Kindred Evil Spirits, Which Attack Mankind*, the title of which is unfortunately more exciting than the actual text. Its author, Dr. Thompson, was a British archaeologist–cum–intelligence officer who would later be stationed in Iraq and reassigned to archaeological duties when that country and all its ancient riches fell into English hands after the First World War. War and wisdom, a single deity. Thompson's writing is as good an example of Orientalist prejudices as one can find, citing anecdotal evidence from contemporary Malaysia, Syria, and Sudan alongside ancient Mesopotamian texts, as if they were all emanations of a single culture of universal primitivism, unvisited by the Western gods of history except as passive objects of observation. I'll come back to him.

But Thompson does talk about Lilith, an otherwise obscure figure who by the turn of the century had already been woven into the new mythologies of modern Europe, the tales Europeans were telling themselves to reconcile themselves to the unprecedented and contradictory realities with which they lived. Specifically, Thompson mentioned two more ancient forms of her name: Lilîtu and Ardat Lilî. He says little of the former, only that she is a "night spirit," and a bit aloof. Ardat Lilî has more intimate relations with humans. She is, Thompson suggests, "a restless ghost that wanders up and down, forced by her desire to roam abroad," bringing illness and misfortune to the men with whom she lies. This is likely anachronistic, a layering on of the Victorian-age preoccupation with Lilith as femme fatale, a demonic incarnation of all ills associated with female desire. Other and more reliable sources suggest that in her earliest

Mesopotamian incarnation it was women who were endangered by this early Lilith, not men, that she winged into houses in the night, causing miscarriages and killing infants in their beds. She was the female spirit on which otherwise incomprehensible evils could be blamed. Her name was invoked on tablets and amulets hung on the walls of homes: "O you who fly in darkened rooms," read one, "Be off with you this instant, this instant, Lilith, thief, breaker of bones."

It is perhaps as this sort of demon that she makes her sole appearance in Hebrew scripture, in Isaiah 34:14, in which the prophet describes the vengeance that God will take on the enemies of Israel. The stars will fall from the heavens, Isaiah promises, and the sky will roll up like a scroll. (To update the metaphor, it will slam shut like a book, or vanish without a sound like a closed tab on your browser.) The land of the Edomites will burn and lie desolate forever. It will be populated by jackals, ostriches, hyenas. Wild goats will bleat at each other in the ruins, and "Liliths will settle, and find for themselves a resting place." Most English-language Bibles translate Lilith's name with other terms: "night birds," "night creatures," "night monsters." The King James Version went with "screech owl" as the closest approximation. Whatever we call her, there she'll be, after the stars fall, among the ruins.

For post-exilic Jews in the first century A.D., residing in what is now Iraq, the already-ancient Lilith persisted as a baby snatcher, and something worse. She appears repeatedly in the Babylonian Talmud, which advises the pious not to sleep alone, lest Lilith slip into their beds and seize them. Archaeologists have unearthed ceramic bowls inscribed with spells in Aramaic to ward off Lilith, "Hag and Snatcher." Around the end of the first millennium, an anonymous satirist (or, more likely, satirists) composed *The Alphabet of Ben Sirach* in Aramaic and in Hebrew. Written in part as a series of lewd and farcical interactions between Ben Sirach, the son of Jeremiah, and the Babylonian king Nebuchadnezzar, it includes a new backstory for Lilith, who appears as the first wife of Adam, made not from his rib but out of earth, just as he was. Immediately they begin to fight. In bed, Lilith wants to be the one on top. So does Adam. It doesn't occur to them that they might enjoy taking turns.

"We are equal to each other inasmuch as we were both created from the earth," Lilith insists. Adam won't hear it. She rebels, flying off through the air and, in her rage, speaking God's forbidden name aloud.

Adam runs to tattle. God takes his side, announcing that all will be forgiven if Lilith submits. If she does not, he threatens, one hundred of her children will be killed every day. Lilith, proud, refuses. "Leave me!" she tells the angels who convey God's offer. "I was created only to cause sickness to infants."

The story was most likely meant as a joke, but the lure of a female demon who could be blamed for all manner of ills was too powerful to laugh off, and it was gathering momentum. This Lilith, the rebellious bride, will show up again in the key medieval texts of Kabbalistic Judaism, shorn of satiric intent. In the Zohar she begins in unity with Adam, prior to the differentiation of male and female. Adam falls asleep, and God hacks the feminine aspect from his side, and "adorn[s] her as they adorn a bride." But Lilith does not want to be wed. She flees. Untamed by the bonds of marriage, she can only do ill: "And she is in the cities of the sea, and she is still trying to harm the sons of the world."

Elsewhere in the Kabbalistic literature she appears as the consort of Samael, the archangel of death also known as Satan, and as a seducer of fallen angels, and of Jacob, to whom she came bedecked with jewels, "her words smooth like oil, her lips beautiful, . . . sweet with all the sweetness of the world." Sweet, at least, until she and Jacob have spent themselves in love and she reveals herself as a fierce warrior "in armor of flashing fire." Elsewhere she is accused of seducing Adam after Abel's death and with his seed bearing all "the

Plagues of Mankind," elsewhere for causing men to ejaculate in their sleep or for scooping spilled semen from the beds of married couples to impregnate herself with ever more demons and plagues. There she will remain, just your average sheet-sniffer, obscure and cast aside like thousands of other forgotten figures of myth, until the nineteenth century when, suddenly, she would become useful again.

❧

It was Christmas yesterday. L. and I went for a hike on a path that looped up through the rocks in an area of the park that was unusually lush with junipers and pinyon pines and even oak trees, the bare rocks heaving with life. At this time last year the mountains to the west, the San Bernardinos, were covered in snow, but they're still bare. The rains haven't come. Not on the coast, where the fires are still burning, and not here. More than three months have passed since the monsoons fell, but the bladderpod bushes were nonetheless in bloom, bursts of brilliant yellow up and down the trail.

For a long time white people didn't think much of this place. In 1853, five years after the United States annexed half of Mexico, U.S. Secretary of War Jefferson Davis, later to become president of the Confederacy, dispatched surveyors to scout out "the most practicable and economic route" for a railroad to the Pacific. The demands of science, conquest, and capital cannot be easily parsed. One of the surveyors, Lieutenant R. S. Williamson of the Corps of Topographical Engineers, wrote that "nothing is known of the country" between the Mojave River and the mountains stretching south and east from the San Bernardinos: "I have never heard of a white man who had penetrated it. I am inclined to the belief that it is barren."

But people lived here then and had for a very long time. Until the 1860s, the area surrounding the desert spring known as the Oasis of Mara was the home of the Serrano people. The Cahuilla ranged through the desert to the south and west, the Chemehuevi and the Mohave to the east. They knew where to find water, and lived well off of jackrabbits, cottontail, bighorn sheep, deer, pinyon nuts, acorns, and mesquite beans. Two years before her death in 2000, a Serrano elder named Dorothy Ramon published a book recording as much as she could of her people's language and traditions. She described a landscape that was anything but barren: "Their Lord was living here, with them, he was alive, not dead. He was like us, alive here. And he would speak to them. He would explain to the people about how to live, about how to get along here on earth . . . He asked them whether they would allow themselves to be transformed to make medicine, so that medicinal plants would grow." Some people became plants. Others, at the request of their god, became deer.

The Serrano creation epic, like the K'iche' Maya's, involves two twins, Pakrokitat and Kukitat. In a version told in the early twentieth century by an elder named Benjamin Morongo, then eighty years old, to the anthropologist John Alden Mason, Pakrokitat labored to create the first humans, but Kukitat, ever mischievous, didn't like the way they looked. He thought they should have hands like duck feet and eyes and bellies in both front and back, and that they should die. The brothers quarreled, and Pakrokitat decided to leave, to create another world that would know neither death nor decay. Kukitat kept this one and lived on among the people, inciting them to fight one another until they grew tired of his taste for destruction and conspired with a frog to poison him. When Kukitat died, they burned his body, but it was too late. The people kept fighting among themselves, as they had when Kukitat lived.

Dorothy Ramon recorded a different story. Despite her efforts to preserve it, she was the last fluent speaker of Serrano, the last person on earth to think and dream in a language that had once been spoken from Los Angeles County almost to the Nevada line. The tale she told

involved another world, a planet, once bountiful, that had been ruined and exhausted. The Serrano, according to Ramon, "used to live
somewhere else. They were living on some planet similar to this one."
It got too crowded, and the crowding caused trouble. People began
killing one another, so "their Lord brought them to a new world . . .
This was to become the new planet. It was a very beautiful world. The
Serrano talk about this in their songs . . . Coming from that other
planet they started over," at the oasis they called Mara.

It didn't last. Worlds die all the time, and new worlds are born.
By the early 1860s, the Serrano had left the oasis. Most historians
blame smallpox. Ramon grew up with another version: White people arrived and "hunted them. They did all kinds of things to them.
They killed a great many of them. They were lost." Most of the survivors moved about fifty miles to the southwest, to the Morongo reservation at the foot of the San Bernardino Mountains, which was by
then functioning as a catchall refugee camp for the displaced tribes
of the Southern California desert: Serrano, Cahuilla, Chemehuevi,
Cupeño, and Luiseño. Ramon was born there in 1909.

The few Serrano who remained at the oasis were soon joined by
the Chemehuevi, who had been displaced in a war with the Mohave,
who were in turn being displaced by white settlers. In 1860, after
several years of resistance, the Mohave had come to terms with the
U.S. Army. According to a Mohave account recorded in 1903, the
Chemehuevi, with whom the Mohave had lived in peace, continued to sporadically ambush settlers, miners, and soldiers until 1865,
when the Mohave, with the army's encouragement, took up arms
against them. Some fled to the Oasis of Mara and learned what the
Serrano already knew, that whites had begun to settle there too.

These were my people. I don't mean directly—that I know of,
none of my kin made it this far west until later in the twentieth century, but they were of European stock, as I am. They were hungry

people, the landless offspring of generations of the landless and de-spised. Their own gods had been killed off centuries earlier and they were raised, as I was, to regard nothing in the universe as sentient other than themselves and perhaps somewhere an unseeable deity who demanded of them mainly a rigorously abstract faith, which is to say, submission. Surely, though, they felt some awe or fear at the limitlessness of the land, at the absence of anything in it that they could recognize as *past*, and at the enormity of the future bearing down on them, the apparently endless possibility that it contained.

They called the area around the oasis Twentynine Palms, after the trees the Serrano had planted there. Their cattle ate and trampled the plants on which the Serrano and the Chemehuevi relied for food. They shot the wild animals, leaving few for anyone else to hunt. In 1875, the state claimed the land around the oasis and gave it to the Southern Pacific Railroad, which was building a second transcontinental route across the South. The tracks themselves would be laid miles away, but the Southern Pacific wanted the water, and denied the land's earlier inhabitants access to the spring. Nothing says progress like a railroad.

By the early twentieth century, nearly all of the Chemehuevi and Serrano who had not succumbed to malnutrition and disease left the high desert and moved to the Morongo reservation, where an earlier wave of Serrano had taken refuge before them. By then the Southern Pacific tracks had cut the desert in two just down the grade from the reservation. There's a casino there now, with a hotel twenty-seven stories high. You can see it from miles away. At night its LED display is so bright that it's blinding. When L. and I drove back from L.A. last week, it was blaring the words ALL YOU CAN EAT SPAGHETTI AND MEATBALLS for all the desert and the mountains to see. We are still hungry, apparently. The Chemehuevi have their own casino, smaller and blessedly dimmer, on the tiny parcel of land allotted to them in Twentynine Palms.

The park service took over the oasis in 1950. By then the spring had run dry: the water table had fallen as the population grew. Water is flowing there again these days. The park service pumps in thousands of gallons each week to keep the oasis green and the few remaining palms alive.

That conviction—that man alone is sentient and the universe just dead, dumb matter and all other animals little more than animate machines incapable of intelligence or emotion let alone something so precious as "consciousness"—would become a point of pride, a badge indicating the superiority of Europeans over peoples deemed "primitive." In an 1885 lecture entitled "From Savagery to Barbarism," John Wesley Powell, director of the U.S. Geological Survey and of the Smithsonian Institute's Bureau of Ethnology, identified the belief in the vitality of natural and celestial objects as a defining characteristic of "savagery." "Superimposed on this," he said, "is found an exalted conception of the wisdom, skill, and powers of the lower animals."

By the late nineteenth century, this was an established commonplace of Western thought. Progress could be measured as maturity or whiteness, on an axis of time or of space. It amounted to the same thing: an ascent into the solitude of a sterile and lifeless cosmos. Only once we imagined the world as dead could we dedicate ourselves to making it so.

It is also in the nineteenth century that Lilith comes to life. After her appearances in the Zohar, she was allowed to doze for six long centuries before being torn from sleep in 1808, by Goethe. It was hardly worth waking for: he gave her a flashingly brief part in *Faust, Part One*, her first appearance in the secular literature of the West. Or anywhere. Blink and you'd miss her: Mephistopheles takes Faust to a witches' orgy on Walpurgis Night. Among the many wraiths and spirits gathered on the mountaintop, the good doctor notices one. He asks Mephistopheles who she is.

"'Tis Lilith," the devil answers.

"Who?" asks Faust.

Mephistopheles explains: she was Adam's first wife and these days traps young men with the beauty of her hair. That's it. Faust is distracted by the sight of two witches, one of them young and lovely. They join them on the dance floor. All goes well until a red mouse emerges from the mouth of Faust's dancing partner. Faust is sorely discouraged, but by then they have moved on from Lilith.

That one quick mention was enough to thrust Lilith into literary currency—for the next hundred years, few texts would be more widely read than *Faust*. In the middle of the century, she would receive a more expansive and less flattering treatment in Victor Hugo's unfinished epic poem, *La Fin de Satan*. Here she is no longer the mere patron demon of wet dreams and sticky sheets that the Kabbalistic authors made her out to be. She is Satan's daughter and representative on earth, the very embodiment of evil. Hugo fuses her with the Egyptian mother god Isis, goddess of wisdom, mourning, and the moon. "I am Lilith-Isis," she declares, "the world's black soul. / Tremble!" This Lilith is decidedly hideous, a veiled "monster woman that Satan made out of shadow." Hugo blames her for corrupting the world after the cleansing of the flood and for providing the cross on which Christ would be crucified. Worse, he makes her the half sister and archenemy of Liberty, the most beloved virtue of the French.

Satan, with Lilith's help, has nearly triumphed. His only obstacle is France, but France, Hugo crows, is "more than a people. It is soul." France is "Man himself." France is Adam, "chasing Night and Death before him . . . all progress is made with his steps." Lilith, in Hugo's telling, is the tyranny of destiny, "the unknown being, baleful and unlimited / that quivering man names Fate." She is a specter, "the fierce eternal blackness of the nights," the enemy of progress and of freedom. Unfortunately for Lilith, and lucky for the French, Satan is sleepy. While their father slumbers, the angel Liberty destroys her.

A few years later across the English Channel—around the time that the Chemehuevi were forced to take refuge at the Oasis of Mara, when the mills of Manchester were just beginning to burn enough coal to cause global carbon emissions to rise—Dante Gabriel Rossetti was painting Lilith, and writing poems about her, casting her in a more alluring if only mildly less monstrous role. On one canvas he elevated her to the aristocracy, titling her *Lady Lilith*, and depicting her as a contemporary Victorian beauty. (He used his mistress as a model.) She sits combing her long, red hair, her full lips pursed, gazing at her reflection in a mirror. John Collier, Rossetti's younger colleague and fellow member of the Pre-Raphaelite Brotherhood, painted her too, making her a pale and longhaired beauty, at once lethal and voluptuous. He posed her in the thickets of Eden with a shining serpent coiling around her leg and low across her waist, its head resting lovingly above her naked breast.

Why did the poets and artists of nineteenth-century Europe see the need to resurrect her, and to frame her for all the evils of the world? What spurred them to retell the biblical creation myth, to alter it, to draw it into the web of narratives that modernity was weaving around itself? Perhaps there was something that threatened them ("Tremble!"), something in the primitive depths of the past that they needed to summon only so that they might again banish it, punish and destroy it for good and all. Perhaps they intuited that, in the great rush of progress, something had been left out, and not just neglected but trampled and violently suppressed. Something important, crucial even, at once seductive and deadly, or deadly at least to those who attempted to evade it. Perhaps modernity had left an irrational remainder, something that would continue to haunt them.

In Rossetti's poem "Eden Bower," Lilith is easy on the eyes and irredeemably corrupt: "Not a drop of her blood was human," Rossetti wrote. "But she was made like a soft sweet woman." The poem is ambivalent to the core: an indictment of the dangers of female sexuality that is at once intensely and bizarrely erotic. Like Hugo, Rossetti exceeded the Talmudic and Kabbalistic sources, laying the full blame for the fall from paradise in Lilith's demonic hands. This is myth at work: European modernity needed a scapegoat, a receptacle ample enough to hold its own insistent ghosts, the vast and savage violence on which the bright, right-angled future had been built. It needed something that would allow it to shed the irrationality that hounded it, to deny that "the world's black soul," in Hugo's words, was its soul too.

Despite their undying popularity, scapegoats never work. At least not as they're intended to. They do not exculpate the guilty, but compound their guilt, creating the need for additional scapegoats, and further crimes. This can and does go on forever. Perhaps Rossetti intuited this: to pin any blame on Lilith was to condemn himself too, and not just him. So he did her a courtesy that Hugo had withheld: he put much of the poem in Lilith's voice, allowing her to tell her own story, albeit in his words.

Lilith addresses them to the snake, also known as Satan, her lover and her only listener. This puts the reader in an unorthodox position, archfiend as well as audience, seducer and seduced. From there it is possible to empathize with Lilith's rage and at the same time to despise its consequences. Spurned by Adam, she beguiles the serpent into letting her take over its body so that she can tempt Eve and wreak revenge: "Bring thou close thine head till it glisten / Along my breast, and lip me and listen. / Am I sweet, O sweet Snake of Eden?"

She is sweet, and thus does Eden fall. Rossetti could lay it all on Lilith and at the same time implicate his reader. The results are the same: The springs that water the garden go dry. The earth hardens. Abel is born, and Cain. "The sword turns this way and that for ever."

There's a marine base north of the Oasis of Mara. At night, from my driveway, I can see the glow of its lights behind the hills. Early in the Cold War, the marines decided they needed a large and uninhabited area for live-fire artillery training. Almost a century after the Williamson expedition, most Americans still understood the desert primarily as waste: a lifeless land without use unless it could be profitably mined or otherwise destroyed. Hence the Nevada Proving Grounds, where the United States detonated more than nine hundred nuclear bombs between the 1950s and the 1990s, and the White Sands Missile Range in New Mexico, where the first atom bomb was tested on July 16, 1945. And hence the Marine Corps Air Ground Combat Center Twentynine Palms.

The latter, which covers more than nine hundred square miles of desert, contains several mock-Afghan villages and, for urban-warfare training, a prefabricated city the size of downtown San Diego. If I climb the rocks behind my house I can see the fake city, a blur of white structures at the base of a hill across the basin to the south. For years, every marine unit headed for Iraq spent a month in Twentynine Palms. A musician I used to know here paid his bills for a while by hiring himself out as an extra in the training exercises, until the marines, for added verisimilitude, stopped hiring local actors in favor of actual Arabic speakers. Most days I can hear and feel the thudding of the artillery concussions as the soldiers train on the far side of the mountains. The windows shiver in their panes, a reminder that, however peaceful the place may seem, a machine out there is constantly at work, readying itself for slaughter. Some weeks are worse than others and though I know the marines' training schedule follows its own bureaucratic mandates and does not immediately reflect the state of the world outside the base, it's hard not to feel the menace building and imagine that it means something, to try to read the explosions like the stars. This year the marines took Christmas off. I'll take it as a good sign. The desert was still. Even the wind let up.

I tried to stay away from my phone and my laptop but I couldn't help myself and found, among the holiday recipes and end-of-year lists, North Korea condemning new UN sanctions as "an act of war," the English freaking out over Russian warships patrolling just outside British waters, and Russian submarines prowling around the undersea data cables beneath the north Atlantic—one little snip could sever Europe from the internet. The commander of the U.S. Marine Corps visited Norway, which shares a 120-mile border with Russia, to address the marines stationed there. They should be prepared, he warned them, for "a big-ass fight."

"I hope I'm wrong," he said, "but there's a war coming."

❧

I hope I'm wrong, but what I am calling the Time of Crisis, Vertigo Time, etc., is also the time of war. Of the war that hasn't started yet. Or that has without our realizing. Or the war that never ended, to paraphrase the poet Natalie Diaz, and that somehow keeps beginning again. Things start spinning. Events whistle as they circle past. The hiss is deafening. Nothing is new that isn't repeated as catastrophe or caricature or farce. I read today's news last month but this time it's worse. The cycles grow shorter. They accelerate and slow and quicken once more. Things go still for a moment and the tension becomes almost unbearable. When they fly off again it's even harder to take. It can be deafening, a high-pitched whine behind your thoughts. In the night it trills still louder. Like Lilith. And so it goes until everyone—right and left, black and white, pacifists and hawks—wants only for it to stop. So they can rest and hear their thoughts again. So that time can again slink forward. Unhurried, the way it used to. But it won't, because that wish—for something to happen, anything, so long as the tension snaps—is a wish also for war. That is how we get there.

The Mohave too had a goddess of war. She was not so different, in some ways, from Athena. Her name was Nyohaiva. Born when the earth was young and wet still with newness, she wandered from place to place and everywhere she went people called her by a different name. She did not argue with them. She was the same, whatever name they gave her. "I teach you only singing," she said. "I do not tell you what you are to do, but only how to sing."

This was not quite true. She did teach them other things: how to paint their faces and their hair in preparation for battle, how to dance the war dance, and to take the heads of their enemies. She taught by doing, gathering one tribe against another, slicing her enemy's head from his neck with nothing more than her sharpened thumbnail, tying his scalp by the hair to a branch cut from the willows that grow on the sandbars of the Colorado. Perhaps, as she saw it, all of that amounts to song.

"I wish all tribes to fight," Nyohaiva said. "When there is a war and a scalp is taken, people will do as I have done. They will dance and enjoy themselves. All will be happy and play and sing." Then, having delivered her message, she turned herself to stone.

Reginald Campbell Thompson, the aforementioned Orientalist, military intelligence officer, and aspiring demonologist, also wrote about owls. In Arabia, he claimed, they were believed to be bereaved mothers transformed by grief, calling out to their lost children in the desert night. In Madagascar, Malaya, and Sudan, Thompson wrote, their appearance was regarded as an ill omen. The ancient Assyrians reckoned an owl's cry at night to indicate the presence of the Seven Spirits, who, per Thompson, are invoked in many surviving Sumerian incantations and poems and "reappear in various shapes and forms in the legends of other Semitic nations." He quotes an ancient tablet:

> They are the Children of the Underworld . . .
> They are the bitter venom of the gods.
> They are the great storms directed from Heaven,
> They are the owls which hoot over a city.
> Knowing no mercy, they rage against mankind,
> They spill their blood like rain.

This, again, is from Thompson's 1903 work, *The Devils and Evil Spirits of Babylonia*, which consists mainly of his translations of cuneiform texts written in ancient Sumerian, a language that he had not apparently studied. "His interpretation of them is apt to go astray," his colleague G. R. Driver conceded in an otherwise laudatory obituary published in the *Proceedings of the British Academy* in 1944, three years after Thompson's death.

Thompson was, by all accounts, a confident man, and not one to be inconvenienced by ignorance if it happened to be his. He was among that class of hardy Englishmen of good birth who took to the seas in the latter half of the nineteenth century, crossing the Mediterranean and the deserts to the east and south of it by railroad, caravan, and camel, risking death by exposure and disease in search of adventure, ancient wisdom, and whatever plunder could be had along the way. This, at the time, was science. No one called it theft. (Not in English anyway.) If Mesopotamia was, as the cliché has it, the cradle of civilization, its adulthood was without question lodged in London. Unearthing treasures in the deserts of Iraq and shipping them to England was, in a way, sending them home. They had simply been lost beneath the sands, waiting to be reclaimed by their rightful owners. "In the standardized orthodox textbook accounts of Middle Eastern history," writes the scholar Zainab Bahrani, "Sumerian, Babylonian, and Assyrian cultures can have absolutely no connection to the culture of Iraq after the seventh century A.D. Instead, this past is grafted onto the tree of the progress of civilization, a progress that by definition must exclude the East."

Thompson had begun studying the Assyrian cuneiform tablets in the British Museum while still a boy. In 1904, when he was twenty-eight, the trustees of the museum dispatched him to Iran. Specifically to Bisitun, where Darius the Great had a monumental text carved in three languages on a high cliff face above the old caravan route between Baghdad and Tehran. All three languages—Old Persian, Elamite, and Akkadian—had fallen into extinction, but their presence together turned the inscription into a sort of giant Rosetta Stone. Thompson and a more senior Assyriologist jerry-rigged cradles out of packing cases strung from long ropes and had themselves lowered over the edge of the cliff "by natives stationed on the natural ledge above." Hanging precipitously, their lives in the hands of these anonymous and quickly forgotten men, they photographed and copied the texts, which they published upon their return to London.

The names of many of the places Thompson passed en route will be familiar from the news—Deir az-Zour, in Syria, besieged from July 2014 to September 2017 by the armies of the Islamic State; Mount Sinjar, where IS massacred Yazidis by the thousands; Mosul, "liberated" from the Islamic State by U.S. bombers and reduced to rubble in the process, killing thousands more. The foundations for these conflicts were just being laid around the time of Thompson's visit. He stayed in the region for a year, starting his own excavations in Iraq, sending home every artifact he could. In 1911 he returned to Syria, joining T. E. Lawrence, among others, at a dig in Carchemish that appears to have served as cover for a less scholarly mission: the archaeologists were there to spy on the Germans who happened to be building a railroad through the region, connecting Baghdad to Berlin. Nothing says progress like a railroad: the British and the Germans were racing to control the recently discovered, and massive, oil fields of the Middle East, and any potential distribution routes of the newly precious fuel to Europe. The age of oil had not yet begun, but the amount of carbon dioxide in the atmosphere, after a mere half century of largely coal-fueled industrialization, had already jumped by 5 percent.

The war that broke out three years later would be, in the words of historian Timothy Mitchell, "the first great carbon-fueled conflict," in which coal-fired factories enabled the "mass production and deployment of the machinery of death," including battleships that ran on Middle Eastern oil, allowing "European states to sustain a war of attrition that massacred millions." Thompson was one of many British archaeologists, Lawrence the most famous of them, to enlist with military intelligence units. This sort of overlap was not new: the inscription at Bisitun had first been decoded by H. C. Rawlinson, a British army officer and political agent for the East India Company. In practical terms, these stalwart adventurers knew the lands and people of the Ottoman East better than any other Englishmen. The tablets and stelae that they sent back from Mesopotamia were not, for the most part, valuable in themselves. There was no gold on them, no gems, no hidden liquid core of oil. They were just carved stones and slabs of clay pecked with shallow notches. But they were believed to be the oldest written texts on the planet, and hence lay at the very root of "Western Civilization," which, impelled by the great god Progress, moved like the sun from east to west, from the Tigris and Euphrates across the Jordan to the land of Israel, from there to the Greeks and Romans—whose treasures other Englishmen had looted—over the landmass of Europe and across one last lick of water to England. So the story went.

The rocks and hardened clay that Thompson and his colleagues had shipped up the mouth of the Thames were the building blocks of a narrative that was as crucial to empire as any mere material treasure. All those stone panels, bas-reliefs, clay tablets, the enormous sculptured lion wrested from the entrance of Ishtar's temple at Nimrud that now guards the Assyrian galleries in the British Museum—all of it let them draw a straight line from what they understood to be the effective beginning of time, the supposed site of the biblical Garden of Eden and the home of the world's first great civilization, the first one capable of recording its accomplishments for posterity, to their own, in Westminster and Bloomsbury, Oxford and Cambridge. By expropriating these objects and interpreting them, by owning them, they were taking possession of history itself. They were laying claim to time.

In a rare moment of concern for appearances, the head of the British Museum wrote to the War Office's Mark Sykes to express his concern that the English might seem "to be merely plundering the country in the interest of England." This is the same Sykes who would become known for his half of the Sykes-Picot Agreement, the secret pact by which Britain and France agreed to carve up the Ottoman territories between themselves, determining the borders— and the resentments—that still define the modern Middle East. The caliphate envisioned by IS, you will remember, was intended to roll back the national divisions artificially imposed by Sykes-Picot.

In 1914, Thompson was assigned to Indian Expeditionary Force D. Having secured the oil fields of Basra, the troops were ordered to take Baghdad but instead suffered one of the war's most humiliating defeats, at Kut. Tens of thousands of mainly Indian recruits were killed. Thompson saw little combat. He was charged with decrypting intercepted Ottoman and German radio communications, just as he had previously labored at deciphering ancient tablets. Plagued as ever by the restlessness of his intellect, he, Driver wrote, "spent many hours every day unostentatiously interrogating all sorts and conditions of men, sedentary shopkeepers in the bazaars and nomad Arabs from the open country and, whenever the chance came, enemy prisoners in the cages."

At the close of the war, the director of the British Museum, anxious to take control of the region's ancient past, asked the War Office to release men with archaeological backgrounds from their military duties. Thompson was "detailed to undertake a general supervision of antiquities, with power to conduct excavations." Ottoman prisoners of war were put at his disposal as laborers. The military authorities saw to it that the antiquities they unearthed were shipped to England without the trouble of customs inspections.

In the years that followed, Thompson would publish broadly, penning several novels ("healthy robust tales in an Arabian setting") as well as various learned works in the field of Assyriology, and his own translation, in hexameter, of the *Epic of Gilgamesh*. (The author of his obituary judged Thompson's rendering "spirited, though somewhat unpolished." It is almost unreadable.) He returned to Iraq in 1927 to run a dig at Nineveh. In the meantime, Iraq, now a British protectorate, had passed a law restricting the export of antiquities, which, though it was drafted by an Englishwoman, compelled archaeologists to share their takings with local authorities. Thompson wrote reassuringly to the British Museum that he would nonetheless be sure to get "as many tablets as we can." That same year, the Turkish Petroleum Company, which was in fact controlled by the British and had been awarded a generous, seventy-five-year concession from the similarly British-controlled government of Iraq, drilled into what was believed to be the largest oil field in the world just outside Kirkuk, about 170 miles southwest of Thompson's dig. Oil burst with such force from the earth that it rained down on the surrounding fields and villages for days.

When war returned to Europe in 1939, Thompson, too old to fight, joined the civil defense corps. He took command of the river patrol along the upper Thames. Two years into the fighting, his oldest son, a volunteer in the Royal Air Force, was killed in action while returning from a bombing raid. Thompson died six weeks later, of a heart attack, coming off a shift on the river.

Is it possible to write without plunder? Perhaps that is a more important question than whether any of this will last.

The New York Times took note today. Time is making headlines. The paper blamed the Rhino's "tornado of news-making" for scrambling "Americans' grasp of time and memory." Has it really only been a year since Obama stepped down? Who can remember what happened in June, much less in February? The massacre in the Texas church (November 5: 26 dead), the massacre at the country music festival in Las Vegas (October 1: 58 dead, 869 wounded) already almost forgotten, the hurricanes that flooded Houston (August 17–September 1: 82 dead) and hobbled Puerto Rico (September 16–October 3: 2,975 dead), the Nazi march on Charlottesville (August 12: 1 dead), all of it seems like years ago. The Rhino's talent for chaos and distraction—and, though *The Times* did not mention it, the media's eagerness to play along, turning his every tweet into the tornados they blame him for—is surely part of it. He's like a child opening every forbidden cabinet, shaking all the bottles marked with skulls and crossbones, serving it up as tea. But the Rhino can't take credit, even indirectly, for the hurricanes and fires, mass shootings, van attacks in Barcelona, London, Nice, refugees dying by the thousands in the Mediterranean, war and famine spilling everywhere. It would be nice to blame him, cozy even, but the Rhino is only a rhino, and not even really that.

≋

On New Year's Eve, L. and I packed the tent, sleeping bags, food, a thermos of hot coffee, and as much water as we could carry. We left the car on the side of a dirt road deep in the national park, filled out a backcountry camping card, and walked off into the desert. The sun had set hours before but the moon was nearly full and bright enough that our bodies cast shadows, sharp ones, on the earth. After about two miles we set up camp in a clearing between two sudden outcrops of stone. We wouldn't see them in their full weirdness until morning, but they were the otherworldly, rounded Krazy Kat lumps of orange monzogranite that bring visitors here from all around the world. In the moonlight we could make out, a few yards past our campsite, a giant spherical boulder balanced atop a pyramidal wedge of stone.

We ate bread that L. had baked with cheese and good tuna from the can and we stared up at the stars and the moon until we got too cold. It was long before midnight when we retreated to the warmth of the tent. My bladder woke me a few hours later and compelled me to shove my feet into my boots and stagger out into the new year. The moon, still giant, had sunk far to the west. The Dipper, which had been just above the horizon when I went to sleep, was almost overhead. Orion and all the familiar winter constellations had set, but two planets that I hadn't seen for months—Jupiter and Mars— shined fuzzy in the predawn haze, so big and so bright that it was all I could do not to piss on my foot.

≋

I don't think I recorded it a couple of weeks ago when the Rhino's secretary of the interior summoned the superintendent of Joshua Tree National Park to Washington. The park's Twitter account had tweeted out some of the conclusions of the federal government's own National Climate Assessment, which had been released a few days earlier. It was straightforward stuff, a very basic primer on climate change: "An overwhelming consensus—over 97 percent—of climate scientists agree that human activity is the driving force behind today's rate of global temperature increase," one tweet read. Others spoke to the likely effects of warming on specific desert ecosystems. The park's namesake Joshua trees, for instance, are delicate creatures despite their charismatic spikiness. They can survive only at specific altitudes, within a narrow band of temperatures, and depend for pollination on a single species of moth, the larvae of which in turn depend entirely on the tree's seeds. One tweet warned that the trees' range will likely shrink by 90 percent if the climate warms by three degrees. For this indiscretion, the secretary called the park's superintendent to his office in Washington, chewed him out, and sent him on his way. It's as if Nero, bored with fiddling, forbade all mention of flames.

Also the Rhino tweeted that his "Nuclear Button" was "bigger & more powerful" than Kim Jong-un's. If it didn't mean taking the rest of the world with us, I would say that maybe this is how we ought to go out, puffed, bloated, raving, led by a cretinous clown.

It still hasn't rained.

≈

I came across another creation tale credited to the Chemehuevi people, who inhabited the oasis in Twentynine Palms, just east of Joshua Tree, at the same time that Rossetti was painting *Lady Lilith*. It's no stranger, really, than Rossetti's and Hugo's stabs at myth, and in many ways more appealing. Certainly it's funnier. The Chemehuevi divided time into two eras: the one in which they lived and a past inaccessible except through myth, the story time, when "the world was young" and all the creatures that we now recognize as animals were people, just as we are now. I am quoting a woman named Carobeth Laird, who was quoting her second husband, a Chemehuevi man named George Laird.

In her telling of his telling, "At first there was only water," and the earth was made by Ocean Woman, the most ancient and revered of all the Chemehuevi deities who, taking the form of a worm, fell from the sky and sprinkled dirt over the surface of the water. She shaped and stretched the dirt to form land, then scraped *mugre*—dead skin and other grime—from her vagina, and with it made Coyote and his brothers, Wolf and Mountain Lion. Later, because gods can do these things, she took the form of both a mother and a daughter. The daughter, Body Louse, was young and beautiful and almost Lilith-like, but without all that weight of sin. She pranced about wearing only a tiny apron made from the skin of a single jackrabbit. It flapped up and down as she walked, singing as she went: "My jackrabbit apron flaps up and down, flaps up and down."

Poor Coyote couldn't resist her. After a series of trials—among other tricks, she tried to drown him—she finally allowed him to make love to her. This wasn't easy, as her vagina had teeth, but Coyote, being clever, substituted the neck bone of a bighorn sheep for his own more fragile member. Once he had knocked out the teeth, the two were free to go at it all night, and night after night after that. Body Louse's mother removed her daughter's fertilized eggs, tied them into a basket, and instructed Coyote to bring them to his brother Wolf, who lived across the water, so Coyote transformed himself into a water spider and dragged the basket across the sea. By the time he reached the shore, though, his curiosity was too much for him. He untied the basket and took a peek. "Immediately people just boiled out of it." In the end only "a few weaklings and cripples, together with the excrement" remained inside. Wolf took them out and turned them into the Chemehuevi and the other desert tribes. The dregs left at the bottom of the basket became the Europeans, who stayed across the sea.

In other contexts, Coyote's manhood was not so delicate. He is said to have had four penises, the physical one plus three supernatural organs, the largest of which was called Tuguwi'api, or "Sky Penis." It was so big that, in another myth, having killed and skinned two old women from the tribe responsible for his brother's death, Coyote was able to pretend they were still alive by disguising himself with one of their skins and his Sky Penis with the other. It spoke, danced, and fought just as well as he did.

I don't want to think about it, but I'm moving soon. In a week and a day. I got a fellowship in Las Vegas for a semester, at a literary institute attached to UNLV. That part is very good. The leaving Joshua Tree part less so. I can't pay two rents so we're giving up the house. L. will be going back to work soon and I'll be in Vegas on my own. We've been packing all week. Saturday we had a yard sale and managed to get rid of most of the furniture other than the desk I'm writing at, a few chairs, a sofa and a bed. I could hear louder blasts from the base than usual that day, and the unmistakable crackle of heavy machine gun fire. I have to be out of the house on Sunday and have been trying not to get too sad about it. L. keeps telling me that we'll find a way to live here again, but I'm not sure that's true. Never has the future felt more uncertain. For the next five months I'll be in the same desert, at least, a little farther east, with a few more people and lights around.

Yesterday we drove into the park for a run, not far from where we camped on New Year's Eve. It was late when we got started, after four. The trail was long and straight, passing through a wide and sloping valley of Joshua trees and creosote and occasional clusters of boulders. The sky was overcast, the sun peeking out between the low clouds and the high mountains to the west. As we ran we kept startling the birds in the creosote bushes. Little warblers, I couldn't tell what kind. When they took flight, darting low between the bushes and catching the sun on their backs and wings, they looked less like birds and more like shards of light, gleaming fragments of time racing away as we plodded after them.

The Rhino is melting down as usual, his courtiers attacking each other on TV. It's good entertainment for all the people stuck indoors. The entirety of the United States is frozen solid. My mother, who lives near D.C., has been charting the spread of ice on the Potomac out her kitchen window. Yesterday she saw an animal, a fox or a dog, standing in the middle of the river. My father, in rural Connecticut, sounded shaken when I spoke to him on Saturday. Temperatures there had dropped to zero, with strong winds and a foot of snow. A single fallen branch could take out the power lines, and with them his heat and running water. In Australia the freeways are melting in the heat.

≋

Boxes, boxes, boxes. The house is nearly packed. It feels like a lifetime, but it was only a little more than a year ago that I moved here. Before that I lived in Los Angeles for nearly twenty years, not counting a few breaks. My last home there was an old house on the edge of Chinatown, built in the 1890s, ancient by L.A. standards. The floors and ceilings were made of thick redwood planks. There was no heat, the paint was peeling, and the few windows that opened didn't close, but I loved it. I would have stayed forever, but the landlady sold it and the new owner was planning to renovate and then move in, so I had to go. It was about a quarter mile from the spot where the Los Angeles River and the Arroyo Seco converge, where an expedition led by a Catalan officer named Gaspar de Portolá camped one August night in 1769. They were the first Europeans—and Africans, though the latter rarely get much credit—to explore the interior of what is now the state of California. They paused long enough to say mass, admire the wild grapevines they found growing there, and give the river a name that the city still bears: el Rio de Nuestra Señora la Reina de los Ángeles de Porciúncula.

There were people living there, of course, the Tongva. A village of some size stood about a mile away, within sight of the bluff on which the Portolá party camped. It was called Yaangna. The Hollywood Freeway passes over it now. Juan Crespi, the Franciscan monk accompanying the expedition, did not ask the Tongva what they called the river. It's not clear that he spoke with them at all. They were, for the Spanish, outside the web of language and the power that names hold. (Ursula K. Le Guin: "One of our finest methods of organized forgetting is called discovery.") The name Crespi gave the river was taken from a tiny church in central Italy, in Umbria, outside Assisi, where Saint Francis had been visited by a vision of Christ. It seems worth mentioning that the church in question, and hence the river, and later the city, was named for the Virgin Mary, Our Lady Queen of the Angels, who was, Marija Gimbutas would hurry to point out, none other than the Old European goddess made up in Christian drag, or, if you prefer, the Egyptian goddess Isis, whose worship in Rome would be transferred to the mother of Christ. And that worship of the Aztec mother god Tonantzin and of Ixchel, the Maya moon goddess, would after the conquest be folded into a properly Catholic adoration of the Virgin, who is often depicted standing atop a crescent moon.

Grapes grew wild in my backyard. My landlady told me they had been planted at the beginning of the last century by the priests at St. Peter's on Broadway, back when the neighborhood was still Italian. But who knows? Maybe they were the same ones Juan Crespi praised in his diaries. Old vines ran along the fence that separated my driveway from the neighbors' yard. If I didn't trim them back or if I went away for more than a few days in the late spring or early summer, their tendrils would quickly cover my car, winding into the wheel wells and over the mirrors and the wipers. When the grapes ripened I gave away bag after bag to friends and neighbors and to Nancy, the security guard at the office around the block. They were sweet and plump and slid off the tongue.

Behind the house, the yard climbed the base of Radio Hill, a little stump of green space that had been amputated from Elysian Park by the construction of the Pasadena Freeway. People lived up there beneath the oaks and eucalyptus, in tents and under ragged tarps. I would see them pushing shopping carts filled with empty plastic jugs past my house every morning, on their way to fetch water. The grapes spilled over the back fence. They ate them too. Every time I walked up that hill, there were more people living there, new dwellings badly hidden among the trees.

The neighborhood was changing. It had become cool, and hence dangerous to those who lived there. Time was speeding up. Old buildings were being torn down and reborn as high-priced condos. The city was pouring money into the park on the other side of Broadway, a sure sign that developers already had their plans in place. The Spanish conquest or the Anglo-American one replayed as speculation-driven flood. House-flipping conquistadors in Volvos spinning six-guns. Every city I have ever lived in has suffered some version of this fate. Money crushing everything, erasing entire neighborhoods, which nonetheless lived on, fashionably undead, grotesque quotations of themselves. Most of my friends in L.A. are artists, writers, musicians, or activists of one sort or another. The ones who had the resources and foresight to buy a home ten or fifteen years ago are doing okay. The rest are holding on to rent-controlled apartments, terrified that their landlords will sell. L.A. had never been an easy town, but it had never been so hard. It's the same in all the other major cities, in the United States and abroad. They're becoming homes for money in which people cannot live.

So I left. I was sad but I was ready. Worlds end all the time. I was up in Joshua Tree visiting a friend and found a place through Craigslist, a sprawling house on an acre of land, a five-minute walk from the national park boundary. It rented for less than a cramped one-bedroom apartment in L.A. The decision was easy. Just before I moved out, the neighbor asked me why I still bothered to water the grapevine if I was leaving. I didn't know what to say. "It's a living thing," I told her.

In the desert I had more space than I knew what to do with. Inside and outside, and inside my head. I could watch the sun rise from my living room and watch it set from the bedroom. Most days I do both. L. had a different job then and could join me for weeks and sometimes months at a time. We learned to slow down. We learned how distracted we had been. We learned the names of the birds and the plants and the stars. But no escape is possible. Capital shrinks space, compresses time. Joshua Tree might as well be on another planet but it is nonetheless an extension of the L.A. housing market, and of financial capitals continents away. Prices were leaping and long-term rentals disappearing, bought up as investments by people in L.A. and New York and rented out on Airbnb, homes for money in which humans sometimes slept.

A friend told me yesterday that my old house in Chinatown was on the market again. The new owner never moved in and never did the renovations. The house had sat empty since I moved out. She is asking for almost $300,000 more than she paid for it a year and half ago. Whoever buys it will almost certainly knock it down and build something bigger, something they can sell at an even greater profit. It's hard to imagine that the grapevine will survive.

For years I lived a mile or two to the west, in Echo Park, which has been long since decimated by the same rolling tide of amnesiac greed. Once, standing outside a bar on Sunset Boulevard at dusk, I looked up and saw an owl, enormous and white, swooping low across the street before it soared off above the rooftops. I stood beside the bouncer, gawping.

"Does that every night," he said. "Soon as the sun goes down. It's the only thing that keeps me from losing my mind out here."

≈

In the Old Testament, the image of an owl roosting among ruins appears repeatedly as a token of desolation. Owls remain where everything else has been destroyed. They're there in the books of Jeremiah and Zephaniah, and in Micah and Job, where the devastation occurs in the first person. "I will howl like a jackal and moan like an owl," Micah laments. "I have become a brother of jackals, a companion of owls," bemoans Job. Nine chapters after the mention of Lilith, owls make an additional appearance in Isaiah. This time God is speaking through the prophet. "I am doing a new thing!" Isaiah enthuses. And he is. This is the one place in the Old Testament in which owls and the desert are mentioned not as signs of barrenness and punishment, but of the generosity of creation. "Now it springs up; do you not perceive it?" God asks. "The wild animals honor me, the jackals and the owls, because I provide water in the wilderness and streams in the wasteland."

Then there's the 102nd psalm. It is a prayer of woe, for vengeance and resurrection. The author blames his suffering on God—"for you have taken me up and thrown me aside"—but he does not appear to hold a grudge. "Do not hide your face from me," the psalmist begs. "My heart is blighted and withered like grass." Too distressed to eat, he is nothing but bones and groans. We've all been there. "I am like a desert owl," he says, "an owl among the ruins."

The psalm ends with an avowal of faith. Everything will perish, the psalmist avers. Even the stars will die. Like God's old clothes, they'll fade and wear out, to be discarded and replaced. But you, the psalmist says to the divine, "you remain the same."

I only just learned that the armored figure on the Great Seal of the State of California is in fact the goddess Minerva. In other words, Athena, the wise. In most versions she is seated. In one hand she holds a spear, in the other a shield embossed with a female head, swarming with serpents in place of curls. In some versions all you can make out are two snakes, crossed and facing one another like an ouroboros, symbol of eternal recurrence and the cyclical nature of time. In others the snakes have morphed into horns and the head is winged, or has been transformed into an owl. A grizzly bear, strangely diminutive, crouches at Athena's feet. Behind her are the glories of nature, the mountains and a wide river or a bay, and four ships at sail on its waters, symbols of commerce and discovery. A miner labors in the near distance, his pick swung back over his head, standing in for industry and extraction, for the wealth that can be wrestled from the earth. The seal's designers could not in 1849 have appreciated the allegorical weight of that bear, reduced in size as it was. It was originally a symbol of Anglo rule: the white settlers who rebelled against the Mexican government of Alta California in 1846 were known as

Bears; their uprising was called the Bear Flag Revolt. In the quarter century that followed California's annexation to the United States, at least 80 percent of the state's native population would die—to disease and to a swift and systematic genocide. Within three-quarters of a century the grizzly would be hunted to extinction in California—none have been seen here since 1924—but it remains nonetheless the state's official animal and unofficial ghost. It's on the state flag too, a haunting, the fierce and cuddly emblem of all that had to be destroyed for this polity to live.

≈

It finally rained.

≈

Carobeth Laird, born Carobeth Tucker, was in her teens and still bearing her father's surname when she fell for a married man and had a child. She did not write about this relationship in any of her books except to say that she had to leave high school but spent every moment that she could, first in Texas and later in San Diego, in the public library, reading anthropology, paleontology, everything. "I had come to believe that there were those who spent their lives in pursuit of absolute truth," she wrote in *Encounter with an Angry God*, her memoir of her life with the linguist and ethnologist John Peabody Harrington, "and I wanted above everything to belong to that elite band."

Harrington—who whatever his many faults may have been was without question a member of that tribe—arrived in San Diego in 1915 to teach a summer school linguistics class in which Carobeth Tucker had enrolled. She was still nineteen and desperate to break out of the narrow strictures to which life appeared to have consigned her. He was handsome and scowling and already, at thirty-one, "in the grip of his grand obsession, his compulsion to record all that could be recovered of the remnants of the cultures, and most especially the languages, of the Indians of Southern California." She was an excellent student. They married within a year.

The romance was brief. She soon understood that Harrington had little energy or attention to spare for her. The native cultures of the American Southwest were dying off, and dying fast, the languages disappearing, the gods and myths dispersing into the mountains and deserts in which they had been born. "The house is AFIRE," Harrington wrote to an assistant in 1941, "it is BURNING." In he raced, salvaging whatever valuables he could grab. He accomplished more than most mortals could. The archives of his work at the Smithsonian Institution are immense—over one thousand boxes containing hundreds of audio recordings, more than thirty-five hundred photographs, and nearly a million pages of notes, most of them unpublished. It took decades just to catalog it all. The guide to Harrington's papers published by the Smithsonian in 1986 is ten volumes long. (It's online. In volume three I found a photograph of Carobeth Laird. She is standing beside a sad-faced Chumash woman identified as María Solares and shielding her eyes from the sun.) In the end, Harrington documented more than 130 languages, including Chemehuevi, Mohave, Serrano, Paiute, even K'iche'. Many of them are no longer spoken anywhere on this planet. They are preserved only in Harrington's careful records and perhaps in the fading childhood memories of the descendants of their last surviving speakers.

All this labor had a cost. Carobeth Laird bore much of it. "Harrington was a man obsessed and driven," she would write many years later, "and he demanded that I share his obsession at the expense of all normal human relationships, even the most intimate, and all the amenities of life." The young couple lived abstemiously, to be kind about it. Laird recalled Harrington scolding her for discarding eggshells without first scraping out the white with a spoon. They quarreled often, moved a lot, and lived always in harsh conditions. The privations were not just material: Harrington regarded physical comforts and emotions alike as unnecessary and expensive distractions. In Carobeth's telling, he was largely incapable of empathy. He was paranoid, and took baroque measures to conceal his activities, his sources, and even his whereabouts from his employers at the Bureau of American Ethnology in Washington, D.C., writing his notes in a code that even his wife could not decipher. He expected her to cook for him, type for him, drive for him, and, increasingly, as he gained trust in her resilience and the formidable powers of her intellect, to do fieldwork for him too. She traveled on her own, identifying and interviewing sources, freeing him to double his efforts elsewhere.

In 1919, Harrington was called back to Washington, "but he saw no reason why my time should be wasted," Laird wrote. "He needed a complete vocabulary and something on the structure of a language definitely recognized as Uto-Aztecan." He dispatched her to Parker, Arizona, to learn everything she could about the Chemehuevi. She was furious. It was May, nearly summer, and she hated the heat. She arrived "sullen, rebellious, totally lacking in enthusiasm." Almost immediately, she was directed to a man who spoke English and Chemehuevi as well as Spanish and Mohave, and who, just as importantly, did not mind interacting with whites.

His name was George Laird. He was forty-eight, "with a handsome, ageless face wrinkled more by sun and laughter than by the passage of the years." His father was "Scotch with a dash of Cherokee," and his mother Chemehuevi, the daughter of a legendary chief. George Laird had grown up with his mother's people but spent most of his life working for whites as a watchman, miner, cowboy, blacksmith, whatever he could find. He was "remarkably at ease" in both worlds, Carobeth wrote, "firmly rooted in neither, but with deep, psychological ties to a doomed and vanishing past" that tripped him at times into crippling bouts of depression. There's a photo of him too in the Smithsonian's guide to Harrington's papers, six pages after the image of Carobeth. He is wearing a suit, resting his hat on his knee, and smiling. Unmistakably, there is love in his eyes.

"By late July, or early August," Carobeth wrote, "I had become proficient enough to begin writing down the Chemehuevi which George dictated. In September we became lovers. The two events were not unrelated." In October, Harrington wrote to suggest that they join him in Washington to speed the work along. Laird agreed to go. The three of them would live there together that winter in a small and shabby apartment, with George sleeping on a cot in the kitchen a few feet from the bed that Carobeth unhappily shared with Harrington, anguish and desire leaking through the gap between the rooms. Harrington was somehow able to blind himself to what was occurring. In Carobeth's telling, he made no attempt to intervene. He kept to his routine, returning from the Smithsonian in the evenings and working with George till night. "The days," wrote Carobeth, "were ours."

Eventually they left him. "I have no recollection at all of telling Harrington goodbye. I cannot say whether we parted courteously or on bad terms," Carobeth wrote. "Somewhere in my memory a ghostly figure of a man stands looking after me, but that may be sheer imagination." She and George drove to California. Carobeth moved back in with her parents until the divorce came through. George found a hotel room nearby and earned money digging ditches. In the years that followed, they continued to record George's memories of Chemehuevi language and lore, but "there seemed to be little prospect of ever publishing our work, and generally other interests, increasing family cares, and the mounting pressure of poverty turned us from it." George Laird died, of pneumonia, in 1940, in Poway, California.

Had he been born a generation earlier, Carobeth speculated more than once, George might have been a revered shaman or a chief. Had the society in which he did live not been so fettered by its own genocidal urges, he might have been almost anything. He would not, at least, have had to dig ditches to stay alive. Carobeth was not bitter about this, and neither, apparently, was George. Despite the difficulties of their life together, she recalled the years they had as happy ones. The books that she would later write are suffused with tenderness for him and with a grief at his loss that had softened but not lessened with the years. This was a different way of knowing, and of telling.

If he did not love them, John Peabody Harrington was without question fascinated by the people he studied, or at least by the structures of their languages and the parallels, echoes, and overlaps of the stories they told to understand the world. His hunger for knowledge was profound, and destructive in its voraciousness. In a 1941 letter to an assistant whose source, the last living speaker of Lower Chinook, had just suffered a debilitating stroke, Harrington wrote: "JUST KEEP AT THE PURE CHINOOK WITH HIM TILL HE KEEPS DICTATING MORE AND MORE, ANY OLD THING, for he will die and what you don't get now he will die with . . . DON'T TAKE NO . . . pester the life out of him till he finds it easier to dictate than not to dictate and he'll do it just as the easiest way out." Harrington collected and conserved with doggedness and brilliance, but knowledge for him was no less madly acquisitive an urge than Cortés and Alvarado's lust for gold, or Reginald Thompson's for cuneiform tablets. He gave nothing back. Some of the communities whose languages he documented have since discovered his archives and used them to recover some part of what they lost, but this outcome was incidental to Harrington's intentions. He hid and encoded his findings, and rarely published. He was, by Carobeth Laird's judgment, "at heart a racist, a great believer in the doctrine of 'racial purity.'" This does not set him apart from many anthropologists of his era, but it did set the terms of the relationships that structured his work, between enlightened observer and primitive observed, scholar and subject, collector and object of collection. He did not believe that any more equitable exchange was possible or desirable.

Carobeth Laird did not begin to write until nearly three decades after George Laird's death. She remained, she wrote, "for some years in a state of shock." She was eighty when her first book, about her marriage to Harrington, was published by the same tiny press that put out Dorothy Ramon's book about the Serrano. Tom Wolfe praised her work in *Harper's*. Larry McMurtry raved about it in *The Washington Post*. In her next two books, she published the bulk of the Chemehuevi stories that she and George had meticulously recorded along with her observations of the Chemehuevi language, kinship arrangements, and sacred rituals. Despite her lack of formal education, they are works of extraordinary depth and analytical rigor, possessed with a modesty that would be hard to imagine in an academic context. "The anthropologist," Ursula K. Le Guin wrote in one of her novels, "cannot always leave his own shadow out of the picture that he draws." Carobeth Laird did not try. She did not pretend to write from "an impersonal standpoint." What value her writing might have, she suggested, was not its impartiality but the passion that guided it, the love, affection, and respect that she bore for George Laird and for the culture that created him. Much of what she recorded, she admitted, came only from him, and had not been independently verified.

"But does this impair its value?" she asked. "Dreams, fantasies, and childhood memories are the stuff of which legend is formed; and legend is surely as important as fact in revealing the soul of an individual, a people, or an era."

z

The moment the rain begins to fall, the smell of creosote spreads across the desert. The air is crisp and cold now, the mountains white at last with snow. Creosote is a spindly, inelegant plant with silvery, segmented branches and tiny, slightly oily leaves. You can coax out their scent by picking a sprig and crushing the leaves between your fingers if you want, but the oils that coat them are volatile and the merest drizzle, even the moisture in your breath, is enough to excite them into releasing their odor without further mutilation. When it rains here, even when it's pouring, I stand outside and suck it in.

Not everyone is as fond of the smell as I am. L. laughs at me when we're out walking and I lodge a sprig in my mustache so I can sniff it as we hike. A friend visiting last week from the coast cringed when I held some to her nose. "Smells like pee," she said. She wasn't entirely wrong, but she wasn't right either. How do you describe a smell without comparing it to other smells? I want to say creosote smells like the ocean but it doesn't, except in its hint of vastness, and its funk. It is simply what the desert smells like: musky, ancient, sharp, eager, patient, and alive.

One of the oldest living beings on the planet is a creosote bush. After a few decades, the individual branches of any given creosote begin to lose their leaves and die, but new stems, clones of the original, sprout from the roots that extend around the plant in a circle underground. What appears to be a ring of separate bushes is hence a single organism and though any of its visible branches may only be ten or fifty or eighty years old, the plant may have been alive for millennia. The wider the circle, the older the creosote. The one that is believed to be the oldest on earth, which is here in the Mojave, in Lucerne Valley about forty-five miles to the northwest of my house, has been around for nearly twelve thousand years. They are hardier than the Joshua trees, and thrive in the heat. Whatever happens to us, they will almost certainly be here for a while.

This is perhaps a more useful way to think about the shape of time—not as a line or an arrow or a circle or a spiral, but something living, a circle that expands out of sight, invisible roots that grow and grow even as the parts we can see die off. "The world is always new," wrote Ursula K. Le Guin, "however old its roots."

In volume three of his monumental *Black Athena*, Martin Bernal mentioned the surprising discovery, in 1978, at Knossos, of ceramic pots, likely more than thirty-six hundred years old, containing the dismembered remains of children. Their bones were scarred with knife marks. The children had been butchered and, apparently, cooked. Nearby, archaeologists found pottery and amphorae decorated with shields and the head of a Gorgon, images associated with Athena. Others would be more cautious, but for Bernal the find confirmed that "Athena was associated with human, and especially child, sacrifice." So much for our "representative of Law, Liberty, and Reason."

In that volume, which is largely concerned with presenting linguistic evidence for the African and Asian heritage of ancient Greece—Classical Greek, by Bernal's estimation, owes as much as 40 percent of its vocabulary to the Ancient Egyptian and West Semitic languages—Bernal also argued that Athena was a descendant of the Egyptian goddess Nēit and a near cousin to the Canaanite deity 'Anat. All three were powerful figures, bloodthirsty warriors

"of renewable virginity," as Bernal put it. All three were associated
with weaving, and with birds of prey: Nēit with vultures, 'Anat with
eagles, Athena with the owl. Plato and Herodotus both attested to
Athena's identity with Nēit. Inscriptions in Cyprus, where Athena
and 'Anat shared a temple, equated the two deities.

"There are no simple origins," cautioned Bernal. It was never a
question of a direct and singular genetic inheritance, of roots leading
up a trunk and bifurcating into branches. Human history, he suggested, was more like a river, splitting off into tributaries, merging
and diverging again and again. Or perhaps like a crowd, joining
arms and letting go, splitting into smaller groups that at times clasp
hands with one another. Or like creosote, cloning itself for thousands of years, the roots living on and producing new shoots yards
away even as one bush or another died or seemed to die. The worship
of Nēit would be largely subsumed by a cult to Isis, the veneration
of whom would in turn be transferred to the Virgin Mary. 'Anat,
whose worship may have involved the sacrifice of virgins, was in
northwest Syria regarded as the sister and sometimes the consort of
Ba'al, the primary god in the Canaanite pantheon, who was alter-
nately adored and rejected by the ancient Israelites. Archaeological
evidence uncovered on the Nile island of Elephantine, in the contemporary Egyptian city of Aswan, suggests that an isolated community of Jewish mercenaries living there in the fifth century B.C.
worshiped 'Anat as the "Queen of Heaven" and wife of Yahweh, who
was regarded by other contemporary Jews as the one and only god.
Behind and beside any One there is always an other, and behind
each of those wait many more.

Bernal's larger argument was not about 'Anat or Athena or even
language, but about the making of history itself. Which is to say
that it was about time, and how and why we conceive it the way we
do. And about the violence and power locked up in those notions,

a violence that always spills out. *Black Athena*, it should be said, sparked a furious controversy. Many of Bernal's etymological claims have been discredited, but his broader historiographical argument holds up well. Until the early nineteenth century, Bernal proposed, there was little controversy about the debts Greek culture owed to Africa and to points farther east. During the Renaissance revival of Greek thought, "no one questioned the fact that the Greeks had been pupils of the Egyptians." In the early Enlightenment too, European intellectuals were openly fascinated by Egypt.

In the second half of the eighteenth century, though, a new paradigm took hold of Europe. It arose, not coincidentally, as European economies reconfigured themselves around the by-then-steady flow of wealth from the Americas, most of it procured thanks to the labor of African and indigenous slaves. This wealth, and the quicker pace of technological change that followed it, brought an ever-deeper conviction in the superiority of European civilization. All that exploitation had to be justified somehow. Or if not justified, disguised. History was reconceived, in Bernal's words, as "the biography of races," a narrative subject to the laws of something called "progress." This extraordinary notion would prove convenient in more ways than one. As an ideology that put European culture at the pinnacle of human history and consigned everyone else to time's lowland wastes, it functioned at once as an explanation of European dominance and a rationale for the slaughter and pillage on which it depended, and continues to depend.

If Europe represented the mature stage of human development, it would need a lineage. Since the genealogically obsessed tend to favor purity, it would help if its ancestry suffered as little admixture as possible. Greece—the people of which were unquestionably European, famously clever, and passably fair-skinned—would be elevated. Hence the subtitle of volume one of *Black Athena*: "The

Fabrication of Ancient Greece." The Greeks would be reinvented to match the most flattering self-image that Europeans could muster: a people supremely rational and wise, with the stubborn and inborn independence of a noble race, uniquely capable of self-governance, mercilessly strong when justice and necessity demanded it. That this image would not have been recognized by either the Greeks' contemporaries or by the inhabitants of the continents being mowed under by "Western Civilization" is another matter. Plato and Aeschylus became the heritage of the English, the Germans, and far-flung white Americans. Greek joined Latin as an indispensable part of the education of the European elite. Classics emerged as a discipline. By the early nineteenth century, wrote Bernal, it had become "increasingly intolerable that Greece—which was seen by the Romantics not merely as the epitome of Europe but also as its pure childhood—could be the result of the mixture of native Europeans and colonizing Africans and Semites."

Egypt, until recently the object of so much admiration and awe, would be "flung into prehistory to serve as a solid and inert basis for the dynamic development of the superior races, the Aryans and the Semites." Recall Hugo's otherwise bizarre conflation of Lilith with the goddess Isis: Egypt had been by then recast as a stagnant mire of corrupt and monstrous superstition. This is the morass that Lilith would stand in for: the terrifying irrational that the nineteenth century propped up as the foundation on which civilization's anxious heights were built, a demon conjured only so that it could be exiled into the wilds of racial difference and female sexuality. By the turn of the century, Sigmund Freud would be attributing an analogous structure to the human psyche: the rational ego teetering above the dark unconscious, the latter conceived as a roiling stew of archaic myth and violent, primitive desire.

The Semites too would soon be tossed out in a hurry. All

Phoenician and Canaanite influence on Greece, and all evidence of substantive cultural exchange with the cultures of Mesopotamia and the Levant, which had until recently been considered uncontroversially obvious, would become suspect. Except to the undeniable degree to which Christianity was inconceivable without them, the Jews too would be crowded out. (And Freud among them: the esteemed doctor fled Vienna in 1938, not long after his daughter was arrested by the Gestapo.) Thus was established what the Israeli historian Shlomo Sand calls "the mythology of white continuity." And thus, finally, we can understand the confusion in that 1936 letter from the Assyriologist Sydney Smith to Arthur Cook. Once again: "To establish a firm connection between Athene and the goddess of the plaque," Smith wrote, referring to the carving of the still-unidentified nude, winged Sumerian goddess with her owls and lions, "will it not be necessary to show that the goddess was not originally, as later, representative of Law, Liberty, and Reason, but a local demon who fell upon the transgressor (witting or unwitting)?" Only thanks to such a thorough bleaching could the European present—and the Greek past—have become so unrecognizable even to those who lived it.

One more thing about Athena. Homer customarily followed her name with the adjective *glaukopis*, which is usually translated as either *bright-eyed* or *gray-eyed* and holds a range of meanings from bright and flashing to blue, light gray, or even green. The word cements Athena's association with the little owl—that is the name of the species, not a description of it—that the Greeks called *glaux* in reference to the brightness of its eyes, and that is currently known by the Latin name *Athene noctua*. "Owl-eyed" might thus have been closer to Homer's intention. The point would likely have been lost on most early-twentieth-century lovers of the classics, for whom fair eyes were an indicator of racial superiority. For the Greeks, Bernal wrote, blue eyes were associated with ferocity, and with misfortune. The glass evil-eye amulets still produced and sold around the Mediterranean—I have one hanging by my door—are a light blue inside a darker blue. "The paleness of Athena's eyes," Bernal concluded, would have "added to the terror she inspired."

z

Of course, my eyes are also blue.

z

She didn't record the year, but it must have been late in the autumn of 1848 when Sarah Winnemucca's father accompanied his father, a chief of the Northern Paiute, to fish the Humboldt River in what is now northern Nevada. He came back with news of a curious sight: white settlers living near the dry lake bed of the Humboldt Sink. European colonization had likely been affecting the people of the Great Basin Desert for a century or more, but only indirectly, as more distant tribes acquired guns and horses and began raiding their less fortunate neighbors and selling them as slaves to the Spanish. These whites, though, Winnemucca remembered thirty-five years later, in 1883, "were the first ones my father had seen face to face. He said they were not like 'humans.' They were more like owls than any thing else. They had hair on their faces, and had white eyes, and looked beautiful."

The first time she saw them herself, she did not find them beautiful. "I only saw their big white eyes," she wrote, "and I thought their faces were all hair." She hid, terrified, first beneath a pile of robes, then behind her mother's back. "Oh, mother," she cried. "The owls!"

z

Last night I stayed up reading another Ursula K. Le Guin novel, *The Left Hand of Darkness*. This one was set on a wintry planet that worships a god who is understood to occupy "the Center of Time," in which no time exists and all times do: past, present, and future, all ends and all beginnings, a great and radical simultaneity from which everything is spun. In the eye of this god, Le Guin writes, "are all the stars, and the darknesses between the stars: and all are bright."

What caught my attention was this line, spoken by an envoy from a future planet Earth: "It is not altogether a bad thing to have criminal ancestors. An arsonist grandfather may bequeath one a nose for smelling smoke."

That seems a little easy, but maybe so. Unless he burns everything down.

≋

I made myself get up before dawn this morning so that I could watch the sunrise through the bedroom window one last time. It was a particularly gorgeous one, the sky streaked with dark, blue-gray clouds lit pink and orange as the sun came up behind the rocks. L. left a few days ago and I've spent the last week engaged almost entirely with objects—which to be sold, which to be stored, which to be given away? Which can I fit in my car and into the small studio apartment waiting for me in Las Vegas? Yesterday I made sure to give myself time for one last walk.

I headed up the wash where K. and A. and I had encountered the owls. It's hard to believe that was just two months ago, a little more. I felt sure I would see them again. When I reached the willows I scoured the rocks. I saw a few phainopeplas guarding their crops of mistletoe. (Phainopeplas live entirely on the berries of the desert mistletoe, a parasite that grows in the branches of cat's-claw acacias. If you watch one for a little while, you'll see it fly a circuit from one cat's-claw to the next, eating mistletoe when it's present, shitting out the seeds when it's not, helping the mistletoe to spread to new acacias. The birds are *farming*.) Quail fled before me like nervous old ladies dressed for church, their little plumed heads bobbing as they scuttled off into the brush. No owls. I even hooted. It echoed a little, but nothing answered.

The wash rose into a canyon with steep stone walls. I paused and sat on a boulder and stared at the lichens on the north-facing rocks. Even in the dim, afternoon light it was an almost neon yellow-green. I think it was A. who told me that it can take one hundred years for lichens to cover a single inch of stone. The patch I was staring at might easily have started growing before my ancestors arrived on this continent, before the Serrano fled the Oasis of Mara, before Juan Crespi renamed the river that flowed past my old house, before Pedro de Alvarado marched on Guatemala.

It was quiet in that canyon. There was no wind and no birdsong, no buzzing bees. Until I heard, above the wall behind my back—it couldn't have been more than fifty yards away—barking, loud, aggressive, fearful. Like a cornered dog rearing up. And then a response, not a bark but a yowl, high-pitched and frightened or in pain. Coyotes, two of them. My first thought was that they were confronting some larger predator, perhaps a mountain lion, that it had injured one of them and that the other, terrified, was attempting to scare it off. But it continued, first the barking, then the yowl, in conversation. Sometimes they would pause for a moment and then one of them would pick it up again and the other would join in. Perhaps there was no mountain lion, just two coyotes fighting over a kill. Or over something of concern to coyotes. Maybe one was injured and the other scared, and calling for the rest of the pack. They didn't come.

At dusk and in the night, coyotes usually go silent after a minute or two. These ones didn't. Three doves flew by in a panic and landed on a rock above me. Eventually they calmed, and cooed. Five minutes passed, ten. The barking continued, and the yowling. Despite the high walls of the canyon I could tell the sun had set. It hadn't gotten darker but the clouds above me had gone pink. Somewhere out of sight, the drama continued. The barking and howling gradually grew softer, losing urgency, until finally they stopped.

I walked back in the fading light. Tiny bats flickered through the air above my head. The willows, when I reached them, were thick with birdsong and the flutter of invisible wings, songbirds bedding down for the night. The walls of the canyon receded. I was in the open wash again, trying not to crash into a cholla or catch my arm on the barbed thorns of the cat's-claws in the dim when I saw it, just ahead of me, silhouetted against the sky. An owl, wheeling low in utter silence, searching for prey on the desert floor.

PART TWO

LAS
VEGAS

4.

I packed the car and drove out east through Twentynine Palms with its marine barber shops and sad bars and massage parlors and I kept driving into the wide and unvaried basin that hangs between the Pinto and Sheep Hole Mountains and that some clever soul thought to name Wonder Valley. Despite recent upticks in its hipness quotient, Wonder Valley remains nearly as sparsely inhabited as it was in the late seventies, when two jets from the base missed their targets and dropped no fewer than thirty-two five-hundred-pound bombs there, causing precisely zero casualties. I headed north over the Sheep Holes and kept going, down into the dry bed of Bristol Lake gleaming white with salt, past the evaporation canals for the salt mines there, the low cone of a long-extinct volcano hulking in the distance. I stopped for a soda in the near-abandoned town of Amboy and asked for ice, but the old man behind the counter at the gas station said he had none because there was no water there. I kept driving, straight through the Mojave National Preserve, past the high, white, almost Saharan dunes in Kelso and through a wide forest of Joshua trees, threadier and taller than the ones I've been familiar with. The sky was enormous and striped with low clouds, the narrow highway spooling off before me as I drove.

Just before the Nevada line, I hit Interstate 15 and almost crashed the car. The sun had disappeared behind the hills and north of the highway I could make out a sprawling grid of solar panels, their right angles jarring but recognizable, at least. Beside them was something I could not understand: three high and windowless black structures, like watchtowers in a concentration camp built by aliens, or for aliens. The towers were surrounded by white, amorphous shapes, low to the ground. They were glowing, emitting an almost blinding white light. Later I did some googling and learned that it was a "solar-thermal plant": a vast array of mirrors arranged in concentric circles, directing concentrated beams of sunlight at the towers to boil the water within them. The white shapes were the reflections of clouds sitting on the surface of the desert. The blinding light was the sun. It had already set, but the mirrors, at a higher altitude than the highway, were still catching its rays.

Across the Nevada line, I passed an outlet mall and two casinos amid the creosote scrub. One was done up in Old West kitsch, the other in turreted Disneyesque fairy-tale medieval. A monorail connected them, and a roller coaster looped between the hotel towers and an artificial mountain. Real mountains looked on from the horizon on all sides. Behind the casinos I could just make out a sprawl of pink stone buildings and concrete guard towers blurred behind barbed wire and chain link. A prison. Finally Las Vegas took shape over the horizon, announcing itself in golden lights and billboards advertising more casinos, marijuana dispensaries, housing developments (THE ULTIMATE MASTER-PLANNED COMMUNITY), Larry Flynt's Hustler Club and various competing strip joints, a racetrack where you could pay to DRIVE YOUR DREAM CAR TODAY, personal injury lawyers, erectile dysfunction therapy (PERFORM AT YOUR BEST), a shooting range where you could fire a .50-caliber machine gun for twenty-nine dollars.

It was dark by the time I reached the Strip and the casinos and the golden megalith of the Rhino's local hotel. The GPS guided me through miles of industrial neighborhoods, or what had once been industrial neighborhoods, the factories and warehouses converted into weed shops and smoke shops and gun shops, twenty-four-hour liquor stores, tattoo parlors, massage parlors, everything lit in screaming neon. Eventually I found the apartment I had rented sight unseen, a furnished studio in a gated complex on a desolate block of the city's downtown. It had its own little yard, a perfect individualized square of Astroturf littered with drying dog shit. The apartment was tiny and brightly lit. I wanted to laugh, but I didn't have it in me: On a low and flimsy bookcase by the window sat the only extraneous objects in the apartment. There was a small pot of molded rubber succulents, a digital clock, and a chubby little statuette of an owl.

❧

All night I kept waking up thinking the sun was rising because light was streaming in through the blinds. It was only the streetlamps. When I was sure that it was day, I got out of bed and went for a run. In the daylight everything seemed washed out, like an old, sepia-toned photograph. The streets were nearly empty, the sidewalks too. The few people I encountered all looked homeless or lost. I ran past the old motels and casinos on Fremont Street and paused to watch a crane demolishing an old hotel as other machines somewhere out of sight sprayed jets of water that spurted up into the wreckage seven and eight stories high. There were still curtains in the windows. I ran past city hall and a courthouse and the headquarters for the city police. I had to stop at almost every corner to wait for the light to turn. At one of them a man spoke to me. He was maybe twenty-five and had long hair, a hoop hanging from his septum, and a stunned look in his eyes. His voice was flat. He wasn't looking at me but there was no one else around.

"It's all fun and games till you hit a red light," he said.

I nodded in vague agreement. The light turned green, but still he didn't move.

❧

In his autobiography, the Romanian scholar of religions Mircea Eliade recalled creeping out of his room as a small boy on a drowsy summer afternoon while the rest of his family was napping. He crawled into the drawing room, which was usually locked and which he was forbidden to enter. "The next moment I was transfixed with emotion," Eliade wrote. "The room was pervaded by an eerie iridescent light. It was as though I were suddenly enclosed within a huge grape." Later in life he was able to return at will to "that green fairyland," he wrote, "and I would rediscover that beatitude all over again; I would relive with the same intensity the moment when I had stumbled into that paradise of incomparable light . . . I would slip into it as into a fragment of time devoid of duration—without beginning and without end." He found it waiting for him even during a prolonged depression in his late adolescence, but by then the experience had become a source of sadness rather than comfort: "By this time I knew the world to which the drawing room belonged—with the green velvet curtain, the carpet on which I had crept on hands and knees, and the matchless light—was a world forever lost."

This "epiphanic vision," as he called it, and his nostalgia at its loss, would sit at the core of Eliade's thinking. There was the sacred—reality stripped bare, the pure, dazzling ontological foundation of the world—and there was the profane, the workaday realm of the relative, the rational and geometric, time that could be chopped and segmented. An abyss lay between them. "Primitive" man and spiritual adepts of past eras had perfected various means of crossing it, Eliade believed, but by the twentieth century humanity had tumbled into a thoroughly "desacralized world." Secular modernity had left us with a cosmos that was *completely* profane."

The sharpness with which he held on to these distinctions—sacred and profane, pure and impure—defined his politics as well. As a young man in the 1930s Eliade had allied himself with the ultranationalist and violently anti-Semitic Iron Guard, Romania's creepily spiritualized analog to Germany's Nazis and Mussolini's Black Shirts. The local fascist movement was "so profoundly mystical," Eliade had gushed, "that its success would designate the victory of the Christian spirit in Europe." In a journal he kept during the war, Eliade recorded his growing depression as the Allies triumphed and Hitler suffered loss after loss.

Not long after Hitler's final defeat, Eliade began working on the book that would make his reputation in the English-speaking world. *Cosmos and History: The Myth of the Eternal Return*, first published in French in 1949, would allow him to reinvent himself as an apolitical scholar concerned with universal themes. In it, Eliade argued that until the arrival of the Hebrew prophets, time was universally understood as cyclical, and was bound through ritual to the sacred. Regularly recurring religious ceremonies enacted, and reenacted, the creation of the cosmos, allowing their participants to play a direct role in the "regeneration of the world," "projecting" themselves into "mythical time." Only after the prophets, for Eliade, did history enter the picture. In their writings, the defeats and humiliations of the Jews "clearly appeared as punishments inflicted by the Lord in return for the impiousness of Israel." God became an actor in history whose interventions could be tracked over time. Historical events took on spiritual significance. "It may . . . be said with truth that the Hebrews were the first to discover the meaning of history as the epiphany of God."

Mythical time was then replaced by messianic time, which was necessarily linear: it led somewhere. The regeneration of the cosmos no longer recurred in cycles according to a sacred calendar, but was deferred into the future. It would come when God willed it. Time had a beginning and an end, and through them, a meaning. By the twentieth century, deprived of God and the possibility of future redemption but stuck nonetheless on this now single-tracked time, humanity faced a new challenge: how to "tolerate" history, as Eliade put it, and its horrors. "We should wish to know," he wrote, "how it would be possible to tolerate, and to justify, the sufferings and annihilation of so many peoples who suffer and are annihilated for the simple reason that their geographical situation sets them in the pathway of history; that they are neighbors of empires in a state of permanent expansion." He was not

talking about the Jews, but about Romania, which had been occupied by the Soviet Union in 1944 and had, in the peace that
followed, lost much of its territory to that still-powerful neighbor.

Eliade would spend much of the rest of his life downplaying,
evading, and denying the intellectual attachments of his youth, but
it has not been lost on his critics that, even after his reinvention
of himself as a sober universalist, he lay it at the feet of the Jewish
prophets that mankind had been thrown into "the terror of history."
It is indeed terrifying, and likely was for Eliade. His own past had
become a nightmare, at once because he suddenly found himself ob-
ligated to hide it, to disown his deep politico-spiritual convictions,
and because it was gone. The green-curtained drawing room with its
soft, Edenic light was gone. His homeland, lost to communism and
to which he would never return, was gone. The spiritualized future
of which he had dreamed, in which Romania, and Europe, would
find redemption by reclaiming its ancient values, that too was gone.

Eliade's losses were almost infinitely less significant than those
of many who survived the war, but it is perhaps because he endured
them that he was able to form the kernel of an insight that is worth
preserving, and developing: that there is a causal link between the
experience of catastrophe and a messianic, and hence linear, experience of time. Messianism is knotted tight to collective trauma, its
frequent product and near constant adornment. What help are cyclical recurrences when the past offers nothing but pain? The circle
must be broken, if only to escape, to run in as straight a line as
possible toward some other future, the dream of a world made new.

Eliade wasted little ink on the sufferings of the Jews either in the
recent or the ancient past, but should we be surprised that a people
who were slaughtered and scattered again and again should have
been so well able to articulate the desperate hope that all of this suffering might lead somewhere, that it might end with the restoration

of what has been lost? In the sixth century B.C., Jeremiah described Nebuchadnezzar's siege of Jerusalem, the burning of the city, the destruction of its walls, and the slaughter of its nobles. On the ashes of the temple, the prophet erected a vision of redemption: The exile will end, the past will end, the future will bring vengeance. Babylon too will be conquered. The destroyer will be destroyed, all its warriors slain, its treasures plundered, its waters gone dry. Nothing will be left of it but waste and ruins, "and the owls shall dwell therein."

☙

It was probably five when I woke but I managed to stay in bed until the sun was almost up. I pulled on a pair of shorts and a hoodie and jogged north, skirting downtown and the old casino district, J. Cole singing through my earbuds about keeping his faith strong. Yeah, yeah. I headed for what had looked on Google Maps like a park. It turned out to be a cemetery, the grass brown, graves bright with plastic flowers. For blocks the sidewalk outside the fence was lined with the tents of the homeless. A place for the living, a place for the dead. In the distance to the west I could see the mountains, clouds gathering behind them like higher mountains still.

Eventually I circled back and headed home, the sun on my left. From the parking lot of an evangelical church that appeared to have once been a bank, drive-through teller window still intact, I could see the city laid out beneath me to the north and the east. It went on, flat and mainly treeless, for miles until, just shy of the mountains, it stopped and suddenly became desert again.

☙

I keep telling myself that it's the same desert, only paved. As soon as you get outside the city, the same plants are growing, the same animals racing along trails their ancestors scratched into the surface of the earth, the quickest transit from bush to rock to bush. When I stand outside the apartment at night—I stand out there a lot on that stupid square of Astroturf—the sky is so hazy and fouled with artificial light that I can make out only the brightest stars. Every lot and vacant acre and the bare strips along the sidewalks, all of it has been methodically denuded. As if naked dirt were preferable to the undisciplined sprawl of life. The medians that divide the freeways are covered in gravel to prevent anything alive from sprouting forth. In some spots the city or the county or the highway authority has erected metal sculptures of desert plants—rebar ocotillos and Joshua trees—to replace whatever was once growing there.

I pinned a sprig of creosote above the door to remind myself. Most of the time it doesn't work. Only at dawn and at sunset, when the sky puts on the same familiar show—sometimes gaudy, sometimes restrained—can I convince myself with any certainty that this is desert here.

What if time as we understand it—this infinitely segmentable line stretching from unseeable past to unforeseeable future—is not an arrow but a scar? What if, as Eliade suggests, we owe our conception of linear time to the sufferings of a then-obscure monotheistic tribe, torn from its homeland in the sixth century B.C.? Could time, as it clicks away, still carry those traumas with it? And us with them? Could all the divisions to which we subject it—those measurable as minutes and seconds, and the vaguer and more intimate fragments that we call moments—echo an original mutilation that was not metaphorical at all? Could time itself be haunted? Does some basic violence reach into the marching of the hours?

If it does, we don't need to blame Nebuchadnezzar and Jeremiah. We don't need a single point of origin. This haunting finds room for everyone. Catastrophe, at least, is eternal. But for the sake of conjecture, we can tell a story here, a story about time. I have suggested that all narratives are lies, paths clumsily hacked through the knotted snarl of truth. This does not mean that we don't need them, and the fact that we need them does not mean that we should not strive to do without them, to look always for other paths, or no path at all, but a bed, if we can bear it, at the heart of the tangle, thorns and pokey bits and all. But I'm not there yet. I need a way to think through time. So let's begin where Eliade left off, with the Babylonian exile—or population transfer, as today's politicians would call it—and with those nags, the prophets. After a long siege, Jerusalem is burned, its people scattered. This part can be verified. Archaeologists have found evidence of a thorough conflagration reaching into almost every corner of the city. This was in 586 B.C. Thousands had already been deported to Babylon after another siege eleven years earlier. Many survived. Some stayed. Others eventually returned, but their sense of time was almost certainly altered.

Trauma stops time. That's what it does. Catastrophe breaks all cycles. Whatever rhythm had once been attained collapses. This is as true of car wrecks and heartbreak as it is of genocidal wars. If you manage to survive, time starts over. It has to. But it resets. A gulf separates you from what came before. That past belongs to someone else. Time begins now with the disaster. *You* begin with the disaster. What came before is irrecoverable—catastrophe has cut it off—so time starts afresh at that traumatic moment and proceeds . . . into what? We don't know. If we knew once, or thought we did, we don't anymore. The future we thought was promised us belongs also to the past. Time proceeds into the unknown. An arrow shot into the void. Or a scar, inching like a worm through the night.

By the beginning of what we now call the first century of the common era, the Jews, with their hope for the Messiah and their by now inherited, traumatized, and hence linear understanding of time, were one of many peoples spread about the shores of the Mediterranean living under Roman rule. This, a military occupation of their ancestral lands, was in itself a trauma, one that echoed others already centuries old. Like all subject peoples, they resented the encroachments of Empire. But Empire is ruthless. It survives by destroying all resistance. When organized rebellion broke out in the province of Judea in the year 66 of the current era, the Romans crushed it with methodical cruelty. By the year 70, Jerusalem had again been razed, its walls smashed, its rebuilt temple destroyed once more. The historian Flavius Josephus, who fought for the Jews and later became a faithful servant of the Roman Empire, estimated that more than a million died in the siege of Jerusalem and that another ninety-seven thousand were enslaved. Even if he was exaggerating by a factor of ten, or even one hundred, the trauma must have been profound.

Catastrophe is eternal. It always comes back. Some fifty years later it happened again. Another revolt, with still more devastating consequences. Hundreds of thousands more were killed, and many others died of hunger and disease. Alongside these fresh disasters, a new religious movement began to spread among the Jews and from them to other colonized and traumatized peoples throughout the Roman Mediterranean: the Messiah had already come, it insisted, but he'd been rejected, executed like a common criminal. With his arrival, time had started over. With his return, time would end. For good this time.

This new religion, Christianity, sought to set itself apart from other sects. It was not bound to a people or a place. It did not require the memorization of arcane rules of behavior, complex ritual proscriptions and codes of purity. Its most energetic publicist, Saul of Tarsus, later known as Paul, ecstatically declared that Christ had freed his followers from the tyranny of the Law: his followers would be justified through faith. And faith, in this context, was primarily a relationship to time, an orientation to the future. Faith was purest expectation. Time, this thing set in motion with creation and reset with Christ's birth and crucifixion, would end with his return. The universe would be redeemed. The faithful would be rewarded with the greatest of all conceivable gifts: eternal life, timelessness in the presence of God. Paradise was reimagined as an escape from the trauma of time.

This new sect was so extraordinarily successful that within a few short centuries—which would henceforth be counted from the birth of the Messiah—it had been adopted by the Empire itself. A creed of unalloyed expectation, with a few colorful and blood-drenched strings attached, it would prove a uniquely flexible faith, a source of solace to the dispossessed capable also of offering succor to their conquerors. And—this is perhaps its greatest strength—to whatever fresh populations they chose to conquer and dispossess. As the centuries gathered it would be spread by sword and cannonade to the Americas, to Africa and Asia and farthest Australia. The people of those continents—their own calendars destroyed, their songs and feast days banned, their sacred texts burned, their elders and their infants slaughtered, and their ancestors defiled—would in turn be thrust reeling into traumatized time. The cycles were broken, the rhythms destroyed. For them too, time became a wound. But the conquerors were merciful: they brought a salve, a promise that one day this vale of suffering could be escaped. And until that day the conquered could find in Christ what so many others had found before them, the sweet and tearful solace of purest expectation.

This is just a story that I'm telling, but it works, I think. To understand time as I was brought up to understand it, as most of us are, is to contemplate emptiness and loss. In what medium does the present float before it becomes the past? Not in fullness but a void. And what is time but constant sundering, loss after loss, a story trauma tells itself about itself? How could it be anything but painful to conceive of our lives as segments, amputated on both ends, adrift in a sea of nothing? And how else might we begin to understand the strange and creosotal growth by which songs about a distant and hazily remembered river, the Jordan, composed more than two thousand years ago by Jews exiled to the banks of another river, the Euphrates, would be sung in a new language on the banks of yet another river, the Mississippi, by the children and grandchildren of Africans torn from their homes and transported to a strange and hostile continent, beseeching the god who appeared to have abandoned them for vengeance and for mercy, to help them back across that river, and into the promised land?

My other grandparents—not the one obsessed with the stars, and his wife, or by that time ex-wife, who preceded him to the grave— were Jews. Which is to say they wore their scars plainly, with hope and a sad sort of pride. They were also communists. Which is to say that they believed in the perfectibility of humankind, in other words in progress, in unalloyed expectation, and that they didn't fuss too much over old scars. One day revolution would wash the tired past away, along with all the injustices and exploitation that framed the world in which they lived. In which we still live. "Reality," as the novelist Roberto Bolaño wrote, "can be pure desire."

I was six when my grandmother died, but I knew my grandfather well, or as well as a very young man can know a very old one. He died at ninety-four, when I was twenty-six. When I visited him, mainly we talked nonsense and made each other laugh. He read two papers over coffee every morning, *The New York Times* and *The*

Philadelphia Inquirer. I remember him folding the sections neatly, patting them down with a sigh as if he were saying good night to a difficult child, and then turning to me, his bright eyes sparkling slyly before either of us said a word.

His father had sold matches on the streets of New York City when he arrived here from Hungary, but my grandfather was born in Philadelphia. He went to university and graduated with a degree in chemistry. That field, he discovered, was closed to Jews, so he found work in a public high school, teaching the subject that he had not been allowed to practice. He met my grandmother there—she was teaching French—and fell at once in love with her and under the sway of her formidable father, who wrote dispatches for the left-wing Yiddish newspapers—first for *Dos Abend Blatt* and later for the *Morgen Freiheit*—that were passed around the tenements and garment factories of lower New York and Philadelphia. They would sit for hours, father and son-in-law, talking and drinking glass after glass of tea. This was the 1930s. Then too, an entire world seemed about to collapse. For many, it did collapse. My grandfather read Marx and came to believe in the possibility of another world, a better one than this.

Years later—years after my grandmother died, years after Joe McCarthy drank himself to death, years after communists were purged from the Philadelphia Teachers Union and my grandfather was locked out of yet another career—he would put pen to paper in an attempt to explain the passions that had guided him. He was in his nineties by then, at the end of his life, but he used the present tense: "I want to grasp the universe and everything in it," he wrote. "I want to understand the nature of the physical world, the dynamics of society, the processes of history, the physiology of the cell and of the whole organism, the neurology and psychology of man, the cultural and individual diversities of all the peoples of the world."

He did not, he clarified, want only to comprehend all this

intellectually, "but feelingly. Such feelingful understanding is akin to love, and so I might say, and do emphatically say, that I want to love. It is this love, I think, not Marxist analysis of class struggle, which has brought me to socialism and beyond."

Often I wonder what he would make of this world, of what has happened since he left it, of the Rhino and the rising oceans and fascism and anti-Semitism okay again as if the twentieth century never happened. Usually I'm glad that he doesn't have to see it. Except that I can still see him, never mind the twenty years that have passed since his funeral. He's right here, time and death be damned. I can smell the soapy smell of him, can feel the crepe-soft skin on the back of his hands, his brittle bones beneath my arm. I can see him leaning forward in his chair to lay the newspapers aside and, as he looks up, before either of us says a word, I can hear the sorrow and the mischief in his laugh.

There is a compressed, ambient violence to Las Vegas that I have never felt in any other city I've spent time in, even places that were actually at war. The boulevards so wide and straight and empty. The supermarkets huge and empty, the twenty-four-hour laundromats empty, the abandoned houses empty, the empty lots empty. Always, and especially at night and early in the morning, a sense that something terrible is about to happen, that it will have to happen, if only to fill all the emptiness. Or that something terrible has already happened without announcing itself, and is happening right now.

I stopped at a thrift store on my way home this evening. I needed a table small enough to fit in the kitchen, something I wouldn't mind leaving behind. But I found nothing, just bright lights and long rows of doleful garments, the stale smell of unloved things. On my way out I noticed that the two guards at the door—one in uniform, the other not—looked suddenly alert. Something was up. They stopped the woman walking out beside me. I kept walking.

I had already started my car when I heard her screaming. I turned off the car and got out. A third guard, in a red and black uniform, had shoved the woman to the ground. He had his knee in her back and was holding a canister of pepper spray to her face. He was telling her to cooperate. "I am cooperating," she was saying. There was panic in her voice. I took out my phone as I walked back across the parking lot. I held it out in front of me and told the guard that I would put him all over the internet if he sprayed her. He was confused, and angry, and we argued for a minute but eventually he holstered the mace, still holding her down with one arm and a knee. He tugged her elbow behind her back and the plainclothes guard helped him drag her to her feet.

"No!" she was shouting, again and again. Then she began to beg. "Please, let me go. Please."

A pair of slacks fell from beneath her skirt onto the concrete. They lay puddled there, beige and limp, almost liquid in their sadness. The other guard pointed them out to me, as if to justify his colleagues' response.

I told him I didn't care what she had done, they shouldn't brutalize her. The guard with the mace took umbrage. He was already leading her off into the parking lot toward his patrol car. She was still begging.

"Brutalize her?" he said, disgusted. "She your wife?"

☙

Walter Benjamin had a very different concept of messianic time than the one I am articulating here, but my goals are not different from his. What I have been calling messianic time—that empty, over-bright hallway with a single door at the end—is closer to his notion of "universal history," a sort of bland and featureless substrate in which events line up, dumbly, "like the beads of a rosary." To this, Benjamin opposed what he called "Messianic time," which he defined as "time filled with the presence of the now." And "now," in his understanding, is not a mere step in that procession, some fragment of a moment that can be marked and measured as it stumbles past. The "time of the now" is pregnant with all that has come before it. The present holds the entirety of the past within it "in an enormous abridgment." And the past, for Benjamin, is hardly dead. It calls out, demanding redress. It exists in the present as demand. It pulses, turning always toward its own redemption.

The arrival of the Messiah—or of the revolution, for Benjamin was a strange sort of Marxist—is a future event to be hoped for and fought for, but not only for the material reversals it will bring about. Perhaps foremost, it is a recovery of the past. It is time healed. Like light through a wire screen, the possibility of redemption surges through the present. The trick is to see it, to hear it, to be able to reveal it, an explosive nucleus of insurrectionary possibility contained in every moment. As Benjamin put it, every second "is the straight gate through which the Messiah might enter."

The struggle in which Benjamin understood himself to be engaged was hence for the past as well as the future, for all those who had been displaced and erased as well as those presently under threat of erasure. It was a battle, defined in terms at once spiritual and political, over history, that it not be conceived via the self-flattering deceits of the victors as a "triumphal procession in which the present rulers step over those who are lying prostrate," in which they succeed at disguising their atrocities as the grand achievements of civilization on the march. "Only that historian will have the gift of fanning the spark of hope in the past who is firmly convinced that *even the dead* will not be safe from the enemy if he wins," wrote Benjamin.

They are still not safe. And as long as they're not, we're not.

≋

I drove back to Joshua Tree for the weekend. I needed some things for the apartment and was happy for the excuse to get out. And I wanted to check on one small thing. So I made that same drive in reverse: out of Vegas, past the Strip and the golden monolith of the Rhino's hotel, past the outlying casinos and the prison and the spooky solar-thermal plant, up through the forest of Joshua trees, past the dunes in Kelso and the volcanic crater in Amboy and the salt flats there and over the mountains into Wonder Valley. I drove past the base, past my old street, and straight to K. and A.'s.

As soon as I got out of the car I had my answer: a week had passed since the rain had come and already the earth was covered with a thin, green fuzz. The grass was coming in. It was all starting over.

✷

The Chemehuevi believed that humans are the offspring of Ocean Woman and Coyote. Not Coyote as in *a* coyote, like the one I watched through the kitchen window of my friend J.'s house in Joshua Tree. He was a big, healthy animal with bright eyes and a thick, clean coat. J. and his partner had constructed a few watering holes around the property, concrete hollows that they filled with water to attract the local critters. The coyote had come to drink but the wind was blowing hard, shrieking and whining, making the creosotes shiver and scattering the smells so that every few seconds the coyote would get spooked by something, trot off in a panic, look around, then slink back for another few sips before scooting off again. It was a fine performance of coyoteness, at once goofy and majestic.

The Chemehuevi were talking about someone else: Coyote with a capital C, one of the Immortals, a godlike predecessor of living coyotes, and of humankind, who bore that name way back in the mythic past, "When the Animals Were People." Until whites began to settle the Mojave in the late nineteenth century, time for the Chemehuevi was split neatly in two. There were present times, in which the stars circled and the seasons of hunting, planting, and sowing rolled past every year, and there were those ancient days, the irrecoverable story times, right out of a dream, hallucinatory, comic and violent and lewd, when Coyote fooled Cicada into sleeping with his own sister, when Owl's wife and son murdered him by trickery only to be killed by Skunk with a cloud of noxious fart, when Coyote ate his own brains and was convinced by his tail to roast himself, when Lizard sodomized Coyote and hid before Coyote could return the favor, when Coyote caught his aunt Bear masturbating with a pestle and she pulled him to her and clawed off his loins and Wolf killed and butchered a mountain sheep so that his brother could be re-loined and Coyote, thus restored, killed his aunt and butchered her and gathered all her insides but the spleen, which slithered from his grasp and spoke: "Tell!" the spleen told him.

"In the story time," George Laird told Carobeth, "everything talked."

That Coyote, the ancient one, our great-grandfather, was a shape-shifter and trickster par excellence, curious and impulsive, lusty and mischievous, a schemer and a fool, "capable of all good and all evil, buffoon and hero, benefactor and villain." He was nothing like his brother. Wolf was handsome, dignified, compassionate, all-knowing. Wolf was "the ideal man." His advice for his brother was always prudent. Coyote always ignored it or, if he on some whim decided to obey, bungled things into catastrophe again. Wolf was wise but humorless. Coyote was a disaster but he knew how to laugh. The Chemehuevi had a saying, at once melancholy and proud, that Laird repeats throughout her books. They could have followed Wolf, they said, but they didn't. "We followed Coyote." It's not clear to me if that "we" refers to the Chemehuevi or to all of humankind, but whatever they meant, it goes for the rest of us too. We followed Coyote.

"That is, in effect," wrote Laird, "the end of the matter."

I couldn't stop myself from looking at my phone before getting out of bed this morning. The board of the Bulletin of Atomic Scientists, I read, moved the Doomsday Clock thirty seconds forward. It is now two minutes to midnight by their reckoning, the latter being nuclear war or some other humanity-imperiling catastrophe. The clock has not been so close to midnight since 1953, after the United States and the Soviet Union both tested their first hydrogen bombs. The board, which includes fifteen Nobel laureates, cited tensions between the United States and North Korea, the United States and Russia, the United States and China, and India and Pakistan; "an insufficient response to climate change"; and "the velocity of technological change," meaning the dominance of easily manipulable digital media, hackable financial and power infrastructures, gene editing, and autonomous weaponry capable of killing without human authorization. The plots of every science fiction novel published since the 1960s are unfurling all at once.

When I got home last night I noticed that a queen-sized mattress had appeared in the empty lot that adjoins my apartment complex. It's a nice mattress, pillow-topped and only mildly stained. I didn't realize in the dark, but I figured out this morning that it was just outside the fence that separates the lot from my own little dog-shit-littered square of Astroturf, concealed by a hedge of bamboo. When I left to go out for a run there was someone sleeping on it, a blanket pulled over his head. Or her head.

I stuck to the side streets to avoid the long red lights and the dazed casino-goers smoking on the sidewalks. The only other people out were the homeless, pushing shopping carts or pulling battered suitcases behind them. A few were ranting, ecstatic or enraged, but most just looked exhausted. I ran north on Main Street, past empty, fenced-off expanses of crumbling asphalt. The buildings that weren't abandoned looked abandoned. (Again, from R. C. Thompson's *Devils and Evil Spirits of Babylonia*: "The occupation of ruins by spectres is a universal superstition." Las Vegas is perhaps not so much a city as a mechanism for turning people into ghosts.) Clouds bunched behind the mountains to the west. The motels had such heavy security screens bolted over their windows that they looked boarded off and dead. I passed an enormous marijuana dispensary and a small sign on the sidewalk identifying the land behind it as the Las Vegas Paiute Colony, a ten-acre plot that until the 1980s was the only land reserved for the people who lived in this basin before whites arrived. There's no sense in building a casino in Las Vegas, so the Paiute opened a twenty-four-hour smoke shop with a drive-through window and, after the legalization of marijuana in Nevada, the dispensary, which is also open all night. The lampposts around it were covered with posters advertising bargain cigarettes. One was for a brand called Timeless Time. It went for $40.89 a carton.

When I got home again a couple, a man and a woman, were lying on the mattress. I can see them now, through my kitchen window on the other side of the bamboo. They're gathering and folding their possessions, getting ready for the day.

$$\approx$$

I almost forgot: earlier this month, on a Saturday, at 8:07 in the morning, a Hawaii state employee chose the wrong option from a drop-down menu and sent a text message to every cell phone in the state, warning residents that a ballistic missile was headed their way. "THIS IS NOT A DRILL," the message read. All through the islands, people said goodbye. With no time for individual confessions, the Bishop of Honolulu, dressed in a T-shirt, offered a general absolution to everyone who made it to church.

I read an article on Twitter with a fresh angle on the scare: the porn aggregator site Pornhub had reported that their local traffic dropped by 77 percent between the moment the alert went out and 8:45 a.m., when the state got around to sending a second text declaring that the first one had been a false alarm. Most people, apparently, didn't want to go out glued to the monitor, with sticky hands and their shorts around their knees. Which is comforting, I guess. By nine o'clock, when the panic had ended and something like relief had begun to sink in, traffic to the site in Hawaii spiked, leaping to 48 percent more visitors than Pornhub would see on a normal Saturday morning. Not dying in a nuclear holocaust is apparently arousing.

❧

The Chemehuevi's neighbors, the Mohave, did not look upon Coyote with such affection. In their pantheon, Coyote functioned as a sort of anti-Prometheus: Tasked with fetching fire to cremate the dead god Matavilya, he dawdled and came back empty-handed. By that time the pyre was already burning—Fly got it going with a stick and dried bark—and Badger and Raccoon and the others were standing around it, weeping and mourning. Coyote cried too, but he wasn't really sad. He was faking. When he saw his chance, he leaped over Badger and Raccoon, stole the dead god's heart before the fire could consume it, and ran off. As soon as the heart had cooled, he gobbled it down. Matavilya's son Mastamho cursed Coyote, condemning him to ignorance, foolishness, and an eternity of wandering.

Unlike the Chemehuevi, the Mohave were a sedentary people. They were agriculturalists who farmed the banks of the Colorado River. It isn't surprising that they saw less to admire in Coyote than the nomadic Chemehuevi did. Coyote's skill at wrestling sustenance from the desert did not make him a role model for them, but a thief, and a competitor for protein. White ranchers would come to hate Coyote too, as much on principle as out of fear for their stock. Over the last century and a half, millions of coyotes have been methodically exterminated, poisoned, and hunted from helicopters and small planes, hung from fence posts as a warning. In Nevada you don't need a license to shoot them. Anyone can hunt them all year round.

But Coyote is clever. When Europeans arrived on this continent, the animals' range was limited to the Southwest and the

Plains. Everywhere else, wolves kept them in check. We were even more methodical at wiping out wolves, though, and over the last century coyotes have spread north and east, into New England, south Florida, New York's Central Park, filling the void left by their brethren. Somewhere along the line they bred with the few surviving wolves and got bigger, stronger. If we followed Coyote, Coyote also followed us, and prospered. He is better at this than we are. And once we're gone—because one day we will go—he will almost certainly remain. With a yowl and a grin, and god's heart smoking in his teeth.

*

If the ratio of people without homes to homes uninhabited by people is a useful index of hidden violence—for what else but the certainty of greater pain keeps people who are cold and exposed from taking warmth and shelter when it's there to be had?—then Vegas is more brutal than most. Last night a police car spent a long time idling beside the lot on the other side of the fence from my apartment. It could have been coincidence, but someone also might have called them about the mattress to report whoever it was for sleeping with more comfort than they were legally entitled to. When I woke this morning there was a man on the mattress. I guess he outwaited the cops. He was alone this time and got up early. He was up already when I went out for a run a few minutes after dawn. I ran almost to the Strip. I made it as far as the Stratosphere, a slender eleven-hundred-foot concrete tower with a spaceship-like disc on top that I can see from my kitchen window. I stopped, stared up at it for a while, felt discouraged, and turned around. When I got back the man on the mattress was gone.

*

Eliade was wrong. The circle and the spiral were not the only shapes that time could take before collective trauma ironed out its curves. It is bigger than that, roomier. There are other ways to see this. Like their neighbors, the Chemehuevi, the Mohave understood time to have been split in two. For them as well, there was the time of the gods, when animals were people and the great figures of myth still walked the earth. And there was the era that followed, in which the Mohave and the Chemehuevi lived, sealed off from their creators by a divide that could be bridged solely through dreams. Until the whites came, that latter era rolled along with the cycles of the seasons. The more important shape, though, was a line: not the one-way arrow of time propelling us ceaselessly into the future, but a horizontal slash that cut a boundary between the gods and us.

This border was permeable in one direction only. The long-past acts of the gods determined everything that happened among humans, but nothing that occurred in this world could influence the gods. Their time was past, their stories over. Dreams, though, could take you to them. "I can go at any time to ask Mastamho whatever I want to know," the shaman Nyavarup told the famed ethnologist A. L. Kroeber. "I could go tonight. It does not take me long to reach him." Dreams were wormholes, paths between dimensions. They made the lost time of the gods almost contemporaneous with us. "The whole basis of knowledge of myth is due to a projection from the present into the era of the first beginnings—is the result of the utter obliteration of time," Kroeber wrote.

Of course, time was not obliterated. It just had a different shape, and different rules. It was neither linear nor strictly sequential. The epic past continues, separated from the present only by our eyelids, by the categories our conscious minds thrust upon the world, and by our skill at dreaming.

A Mohave who fought well in battle, who outwitted his or

her enemies, or performed any task with agility and grace was said to have "dreamed well." A Mohave who stumbled or failed had "dreamed badly." Whatever knowledge and virtues living men and women might carry were on loan from the other side of time and could be borrowed only through dreaming. The patterns that govern life on this side were first established on the other. Many of those patterns, the collective customs and beliefs of the Mohave, are laid out in the story of Mastamho, which was told to Kroeber and other anthropologists by various Mohave shamans—or "doctors," as they called themselves—who insisted they had not heard the story or inherited it from others but had witnessed the narrated events themselves. They had been there, in dreams.

Mastamho's tale begins with the death of his father, Matavilya. The version I will recount was told by another of Kroeber's informants, a man named Jo Nelson, "aged about sixty" when Kroeber interviewed him in 1903, with the help of the translator Jack Jones. Nelson's "mind was orderly, his procedure methodical," Kroeber recorded. "He distinguished between hearsay and actual observation." He cautioned, for instance, that he was not present when Matavilya fell sick, "but dreamed of him and saw him only when he died." He picked up the story from there: Mastamho, still a young boy, and mourning his father, took the opportunity to teach the Mohave how to deal with death.

"When people die I want you to burn them," he said.

He dispatched Badger to dig a hole, Raccoon to bring wood, Coyote to seek out fire. Of course, Coyote flubbed it and got himself cursed. Mastamho went on to finish the work of creation, building a world for the Mohave and showing them how to live in it. He directed the water into springs and a river—the Colorado. He lifted the earth into mountains and assigned territories to the Mohave, the Chemehuevi, and to neighboring tribes. He told them each which foods to eat, what languages to speak. Again and again he warned his listeners, "I will not die like Matavilya, but will become a bird." First, he would tell them everything.

He showed the Mohave how to make sunshades and houses, how to plant seeds on the fertile banks when the river recedes, how to go to war and take slaves, how to make pottery, water jugs, and stewpots. "Do not forget what I tell you," he warned. He delegated Mockingbird and Thrasher to teach the Mohave how to be happy when he was gone, how to feel good, how to marry and love and make children.

Finally Mastamho flapped his arms four times. They turned into wings. He became a bird, not an owl but an eagle. He flew low above the river to the south, where it spills into the Sea of Cortez. There he stayed. He forgot everything he had known, even how to catch fish and how to keep his plumage clean. "He was crazy and full of lice and nits," Nelson told Kroeber. "He is alone, not with other birds, and sits looking down at the water: he is crazy."

Last night I almost crashed the car again. I had to go to an event sponsored by the Institute that brought me here. It was in Henderson, in the southeast corner of the basin. I got on the freeway, the 515 this time. As the entrance ramp curved, it ascended high over the streets and I could see the entire city laid out in front of me, shimmering like a net of gold tossed across the valley floor. At the very center of it were the towers of the hotels and casinos along the Strip, gleaming blue and red and green as well as gold, a strange gem at the center of this still stranger setting. It was almost beautiful. I realized just in time that I was about to hit a concrete wall.

Later, in bed, staring at the glow of the city through the blinds, I couldn't stop thinking about those lights. Why is everyone so afraid of the dark?

☙

A. L. Kroeber was straightforward about the purpose of his work and its intended audience. In the preface to his monumental *Handbook of the Indians of California*, he cautioned that he was not writing "a history in the usual sense," as the "vast bulk of even the significant happenings in the lives of uncivilized tribes are irrecoverable . . . Nor do the careers of savages afford many incidents of sufficient intrinsic importance to make their chronicling worthwhile." Elsewhere, in one of several books on the Mohave, Kroeber wrote, "I have long pondered to whom we owe the saving of human religious and aesthetic achievements such as are recorded here. It is probably not to the group that produced them. Why should we preserve Mohave values when they themselves cannot preserve them, and their descendants will likely be indifferent? It is the future of our own world culture that preservation of these values can enrich, and our ultimate understandings grow wider as well as deeper thereby."

Perhaps it was his own natural sensitivity, or his experience growing up as a Jew in Romania, and later in France, but the ethnologist and psychoanalyst George Devereux was less singularly enthralled with the virtues of "our own world culture" than Kroeber was. Devereux would recall the periods he spent living among the Mohave in the 1930s with such affection and nostalgia that, half a century later, before his death in 1985, he requested that his ashes be transported from Paris to the Mohave reservation in Parker, Arizona, and disposed of there according to Mohave ritual. Nowhere else in the world had Devereux felt so much at ease, and so at home. In the intervening decades he had published dozens of articles on the Mohave and one extraordinary book about them, *Mohave Ethnopsychiatry and Suicide*, which described the Mohave's traditional understandings of mental disturbances in what at the time was an almost revolutionary fashion: on their own terms, as a system that was as accurate in diagnosis and effective in treatment as any devised in the West. The Mohave's attentiveness to dreams and their general sexual openness, he argued, made them uniquely insightful about extrarational states, and empathetic to those who suffered from even the most floridly antisocial psychiatric symptoms. Madness, for the Mohave, carried no more stigma than any other ailment. Devereux's descriptions of the isolation and confinement of the mentally ill in European asylums evoked in them only horror and disgust. They even treated him once, and cured him, of a severe bout of lovesickness.

Regarding Mastamho's final transformation, and derangement, Devereux alluded to the Mohave belief that insanity at times resulted when a person's "knowledge exceeded their heart." When they knew too much, that is, more than they could handle and stay sane. Gods, one would expect, might be particularly susceptible to such disturbance, but Devereux was not willing to develop this idea. "The problem of psychologically disturbed deities," he cautioned, "is an extremely complex subject, which transcends the limitations of the present work."

Paris is flooding. Another new study published last week linked global temperature rise to an increased risk of flooding in central and western Europe. A couple of weeks ago NASA climate scientists ranked 2017 as the second-warmest year on record. The warmest was 2016, the third-warmest 2015. In fact, seventeen of the eighteen warmest years ever recorded have occurred since 2001, i.e., over the last seventeen years. This winter has been Paris's wettest in a half century. The Seine is flowing four meters higher than usual. The Louvre had to close its lower level and hurriedly move thousands of precious artworks from subterranean storerooms. Rats, flushed from the sewers, are racing through the city's plazas and streets. The British tabloids, which have for years been comparing immigrants to vermin, brought the metaphor full circle, announcing an "invasion" of "savage," "disease-carrying" "refugee rats." Cape Town, meanwhile, is poised to run out of water entirely.

≋

Speaking of creosote: the famed anthropologist A. L. Kroeber was the father of the writer Ursula K. Le Guin, whose novels projected encounters between cultures alien to one another onto distant and imaginary planets. Hence that K. ("I don't know what human nature is," Le Guin had a character very much like her father wonder in one of her novels. "Maybe leaving descriptions of what we wipe out is part of human nature.") It was also Kroeber who in 1903 inspired the nineteen-year-old John Peabody Harrington, then enrolled in a summer class at Berkeley, to study Native American languages. Just as Harrington, unwitting, would lead the nineteen-year-old Carobeth Tucker, enrolled in a summer class in San Diego, to a larger life than she had dared to hope for, a larger life than his.

≋

The mattress is gone. No one had been sleeping on it for a few days and I hadn't given it much thought. I guess someone hauled it off. When I went out this morning a man was sleeping in the alley outside the gate, curled up on a square of cardboard that was significantly smaller than he was. I jogged over to Las Vegas Boulevard and headed north, past the cemetery. Everyone there was just waking up, standing bent outside their tents and squeezing their belongings into shopping carts and suitcases, scrounging for the day's first cigarette.

Frank Ocean singing through my headphones about jumping off a roof into a pool. "Kiss the earth that birthed you," he sang.

Las Vegas Boulevard passes through the Strip a few miles farther south. The other night I made a wrong turn and ended up getting sucked right into the heart of it. It was Saturday, and packed. Streams of tourists flowing through the crosswalks from the Venetian to the Mirage, palm trees and palazzos, fountains burbling, everybody jolly. It was early still. I stopped at a red light beneath the Stratosphere, relieved to have reached the end of the Strip, and looked up to see a body tumbling toward me from above. I only noticed the guide wires just before it disappeared behind a wall. I went home and looked it up: for twenty-five dollars you can jump off the thing.

It's not as cold as it was, but there's still snow on Mt. Charleston, which is where, I learned, the Chemehuevi believed that Coyote and Wolf resided. I had to run in and out of the street to avoid the bodies laid out sleeping on the sidewalk. Before I went running I had peeked at the news. The North Koreans told the South Koreans that they were willing to consider giving up their nuclear weapons and would freeze their testing program if the United States entered talks. So there's that. The man was still sleeping in the alley when I got home. He was bearded, maybe ten years older than me. I went inside, grabbed a banana and the leftover chicken I had been planning to eat for lunch, and laid them in a foam takeout carton next to his head. He twitched a bit at the sound of my footsteps, but didn't wake.

❧

The Mohave, by the by, held the conventional and apparently almost universal beliefs about owls, that their presence outside a house was a sure sign that someone within would die.* But they had some other, more complicated notions. For the Mohave, according to Devereux, the ability to cure an illness involved accessing the same power that had caused it to begin with. Shamans could use their talents for good or evil, and benevolent healers had a way of transitioning into malevolent witches. Illnesses and deaths were hence frequently blamed on them. It was easy enough to tell if such suspicions were correct: the heart of a victim of witchcraft, the Mohave believed, would transform into a ball of flame and spontaneously emerge, several days after cremation, from the ashes that remained in the fire pit. It would then shape-shift into what Devereux described as a sort of featherless owl with a human face and ears. Unless the shaman responsible for bewitching the deceased swiftly drowned the awful creature in the river, an owl would swoop down and carry it off to its nest. When night fell, the owl would pop the creature on its back and fly off again, teaching it to cry out the name of whoever it was that bewitched it. Eventually it would turn into an owl itself and would fly through the night on its own, calling out to its killer by name, addressing him or her as "father," so that the witch would be revealed and the victim's loved ones could take revenge and kill the witch in turn.

"Anyhow," said Devereux's interpreter, whom he identified only as E.S., "the owls are Mohave, since they are the hearts of the Mohave."

---

* From a poem by Natalie Diaz, who grew up on the reservation at Fort Mojave:

> Angels don't come to the reservation.
> Bats maybe, or owls, boxy mottled things.
> Coyotes, too. They all mean the same thing—
> death. And death
> eats angels . . .

There was an eclipse last night. This morning, really, about an hour before dawn. It was a blue moon, full for the second time this month. It's hard to believe so little time has passed since L. and I woke up in the desert on New Year's Day. The moon was full then too. It's still January and it's been a long year already.

The Rhino gave his first State of the Union address last night. I couldn't bring myself to watch, but a few hours earlier, at a lunch with various TV news anchors, he bemoaned the "tremendous divisiveness" that afflicts the United States and expressed his desire to unify the country. He cautioned, though, that "without a major event where people pull together, that's hard to do." The assembled journalists did not ask him to elaborate, and the Rhino did not specify what sort of event he meant, except to say that he would prefer to do it "without that major event because usually that major event is not a good thing."

Maybe someone told him about the Reichstag fire.

My alarm was set for five fifteen but I woke up before it rang. I pulled on a sweatshirt and stepped out into the yard. The moon was hanging in the western sky, perhaps a quarter of it still bright and white, the rest a dirty, dusky red. Behind me, in the east, I could see Jupiter rising, fat and blurry in the haze. It was cold out so I got back in bed. But I wanted to see the rest of it and soon I was standing in the yard again, only a sliver of the moon still white. I went back to bed and still couldn't sleep. I got up a third time. The white bit was gone, the moon erased, nothing left of it but a dingy smudge. I didn't want to watch it come back—it could only be disappointing, an ordinary, silvery bright moon again—so I went inside and managed to sleep another hour.

~

Except for that one wrong turn, I haven't yet been to the Strip. I'm sure I'll go, but I'm in no hurry. I went once more than twenty years ago, passing through Las Vegas while driving across the country to move to L.A. It was August. I remember walking for hours from scorching sidewalk to air-conditioned casino to scorching sidewalk again, gawping at the glittering absurdity. Pirate battles staged every hour in a pool-sized sea on Las Vegas Boulevard, the soft carpets of the casino floors, the constant pingpinging of the slots.

I drive past it on the freeway every few days. Driving south from downtown, it begins with the Stratosphere. Then the gleaming, golden cenotaph of the Rhino's hotel, then the Palazzo, the Venetian, the Wynn, and the Mirage, all of which are owned either by the troglodytic Sheldon Adelson or by another friend and confidant of the Rhino who just today stepped down from his post as finance chair of the Republican National Committee after more than one hundred of his employees told *The Wall Street Journal* that he had sexually harassed or assaulted them, sometimes with a German Shepherd in the room that responded only to his commands. In German. Then there's Caesars and the Bellagio and the mini fake New York and the mini fake Paris and the Luxor, a giant black-glass pyramid that looks like the villain's lair from a 1960s sci-fi comic. At night it emits the strongest beam of light in the world straight up from the apex of the pyramid, sending a clear message to any extraterrestrial observers that they need not bother with this planet. On a still night, if you slow down on the freeway you can see clouds of bats diving by the thousands through the beam, picking off the bugs in what amounts to the largest porch light in the solar system. Last is the Mandalay Bay, through the thirty-second-floor gold-leaf-encrusted windows of which Stephen Paddock fired eleven hundred rounds, killing fifty-eight people and injuring more than eight hundred others at the concert across the street. That was in October. It's February, but except for the occasional #VegasStrong billboard, you'd never know this town had just witnessed a massacre. The writer who had this fellowship before me told me that when it happened she felt no surprise. "It sounds awful," she said, "but it felt normal."

For now, downtown is enough. It's the city's past—the graveyard of earlier casinos, their glamor gone sour—and a glimpse at its future. (Benjamin: "We begin to recognize the monuments of the bourgeoisie as ruins even before they have crumbled.") If all of this is screaming one thing, it's not glitz or wealth or even waste, it's transiency. That nothing here can last. And not just here. Vegas is a microcosm of a globalized casino economy, a culture of pure and shiny spectacle, an empire built on buried murders and the adrenaline high of quick and brutal pleasures. Neon flickers on and flickers off again. Dust dulls the sequins and the gold. It's getting hotter and the water's almost gone. Downtown and the Strip are sinking because so much groundwater has been pumped out from beneath the boulevards. The water in Lake Mead is 130 feet lower than it was eighteen years ago. Climate scientists estimate that if carbon emissions are not aggressively reduced, the Southwest faces a 99 percent chance of a "megadrought" one that would kill nearly all of the trees and all of the farms and toss up regular storms of toxic dust that would render the region uninhabitable—before the end of the century.

If there are no clocks in casinos it is not only because ignorance of the passage of time keeps the marks at the slot machines and the house deep in cash. It's so that no one will recall that everything will disappear, that it's disappearing already, that that's what it does. (Benjamin again: "Gambling converts time into a narcotic.") On that first early-morning run I took down Fremont Street last month, a demolition crew was tearing down an old hotel with jets of pressurized water, as if the water wasn't already running out, as if the future would never arrive and the past were a mistake that could be simply rinsed away.

≈

"The wise men tell us that the world is growing happier—that we live longer than did our fathers, have more of comfort and less of toil, fewer wars and discords, and higher hopes and aspirations. So say the wise men; but deep in our own hearts we know that they are wrong." Thus did the ethnologist James Mooney begin his 1896 account of the Ghost Dance religion with a heresy of the highest order. The object of Mooney's study was an outbreak of apocalyptic thirst that briefly, in the early 1890s, united the scattered native tribes of the American West in the belief that the end was coming: the whites would be washed away, the earth rejuvenated and redeemed. This conviction stood directly opposed to the faith common to members of Mooney's own peculiar tribe, the pioneering predecessors of contemporary anthropology: that time was angled upward, that humanity had advanced through the various dusky stages of primitivism to the glories of European-style industrial modernity as if on a staircase, the muddied bottom steps inevitably giving way to gleaming marble, clear glass, and polished steel. (Or perhaps to a black-glass pyramid spitting light into the heavens.) If this conviction was not recognized among Mooney's peers as a matter of faith, no more rational than the expectation that the messiah will appear at 6:00 p.m. next Tuesday, it was only because it was, and in many ways remains, so universally held that it was almost invisible as a potential object of critique.

Mooney himself was ambivalent, or perhaps simply stuck. On the one hand the smooth wall erected by the prevailing certainties of his race; on the other a terrifying landscape of rubble and destruction. He often slid comfortably into the conventional rhetorical modes, equating "civilization" and "the white man" as if there could be no gap between the terms. But he also at times found the courage to describe the entire edifice as a mirage, rejecting not only the fundamental credo that humanity was advancing to greater states of perfection, but the racial hierarchy on which the whole, murderous system depended.

Until his death twenty-five years later, Mooney worked for the U.S. government. As John Peabody Harrington would a generation later, he labored in the employ of the Smithsonian Institution's Bureau of Ethnology, established in 1879 by the same act of Congress that created the U.S Geological Survey. The young nation, having acquired vast lands, desired an accounting of its possessions—be they mineral or human—so that they might be more fruitfully managed. The early Sumerian kingdoms left clay tablets enumerating the possessions of the state: sheep, slaves, grain, taxes owed and taxes paid. The American empire, in its infancy, produced annual reports. The ones published by the Bureau of Ethnology were fat and imposing documents bound in olive-green cloth, stuffed with essays long enough to be books of their own, and introduced by John Wesley Powell, the bureau's architect and champion.

A former Union army officer who directed the bureau from its inception until his death in 1902—he also directed the U.S. Geological Survey for much of that period—Powell described its chief purpose as "the discovery of the relations among the native American tribes, to the end that amicable groups might be gathered on reservations." Only via the fledgling discipline of ethnology, he proposed, could subject populations be controlled and the "Indian problem" solved: if the state did not understand the cultures, laws, and religious beliefs of the peoples it had conquered, it could hardly hope to convince them of the virtues of private property and church on Sundays. Out of this practical goal, early ethnologists almost incidentally, in Powell's telling, came to define "an essentially distinct science, the Science of Man." However far its practitioners may have wandered since, these are the roots of American anthropology.

James Mooney was never entirely down with the program. A child of Irish immigrants, Mooney understood something of the psychic costs of assimilation and the roundabout routes resistance might take. He grew up poor, in a household steeped in Irish folklore and in his mother's recollections of the Great Famine of the 1840s, which wiped out a quarter of the Irish population. In the United States as in Britain, the Irish were still primarily perceived less as a people than as a "problem" to be solved. Newspapers and magazines routinely portrayed them in simian form, as "white negroes," stuck on the lower rungs of the evolutionary ladder. Mooney, who became obsessed with Native American cultures while still a child, surely saw the parallels. In 1890, at twenty-nine, he was en route to Oklahoma, then still Indian Territory, to continue his fieldwork with the Cherokee, about whom he had already published an essay in the bureau's *Seventh Annual Report.* When rumors of a new religion reached Washington, Powell ordered Mooney to bypass the Cherokee and investigate the effects of the Ghost Dance on the tribes of Indian Territory.

The previous year, a Paiute ranch hand in the eastern foothills of the Sierra Nevada mountains had a vision. His name was Wovoka, and he was said to be able to control the weather, to speak all languages and make animals talk, and to render himself invisible to whites. Word spread with astonishing swiftness: a messiah had arisen in the West. Tribes from as far as Oklahoma and the Dakotas had sent emissaries to Wovoka in western Nevada. They came on foot and on horseback and by trains tugged by steam locomotives over nearly fifteen hundred miles of plains, mountains, and deserts and back again, spreading the good news throughout the entirety of the American West. The same railroads that had abetted their dispossession, ending a world, would bring the far-flung seekers to the prophet. The message he gave them would help their people to survive in the new world they had inherited. The details of the creed and its rituals varied from tribe to tribe, but the essentials stayed the same.

A cataclysm was coming. Some thought it would be fire that would cleanse the land of whites. Some thought floods, others tornadoes. It didn't matter: "The white race," Mooney wrote, "being alien and secondary and hardly real, has no part in this scheme . . . and will be left behind with the other things of the world that have served their temporary purpose." In their absence, the earth would be renewed. The sick would be healed and all the Indian dead returned, not only those who had recently fallen but the ancestors, and the animals with whom they had shared the land: "The new earth, with all the resurrected dead from the beginning, and with the buffalo, the elk, and other game upon it, will come from the west and slide over the surface of the present earth, as the right hand might slide over the left."

To hurry it along, Wovoka taught his people and all who came to visit him a dance, which he instructed them to perform together at regular intervals. In some places thousands took part at once, dancing, singing, and "dying"—falling by the hundreds into tearful and ecstatic trances in which they communed with their most beloved dead. "Father, I come," they sang, crying and moaning in grief and expectation. "Mother, I come. Brother, I come." One of Mooney's Arapaho sources showed him a letter dictated by Wovoka that summed up the tenets of the new faith: "Everybody is alive again."

His study of the Ghost Dance would occupy Mooney for much of the next three years. He took part in the dance himself, recorded its songs, photographed its participants. He spent time with the

Arapaho, Cheyenne, Kiowa, Comanche, Apache, Caddo, Wichita, Omaha, Winnebago, Sioux, Shoshone, Navajo, Hopi, and Paiute, traveling, by his estimate, thirty-two thousand miles. In November 1891, he visited the messiah himself just northwest of the Walker River Reservation, near what is now the town of Yerington, Nevada, just under four hundred miles northwest of Las Vegas. He found the prophet hunting rabbits and left some days later weighed down with gifts: rabbit-skin robes, pine nuts, magpie feathers, and a cake of the sacred red paint that Ghost Dancers smeared on their faces to ease communication with the dead.

Wovoka, Mooney learned, had been born in 1856, around the time that white settlement of Northern Paiute lands began in earnest. "They came like a lion, yes, like a roaring lion," Sarah Winnemucca, the Northern Paiute activist and educator, wrote of those early years. Within a decade of his birth, the old ways had ended. Disease, famine, and violence had reduced the Paiute to a tattered remnant. Wovoka grew up in the ruins of the world his parents had known. He found work with a rancher named David Wilson—local whites knew the prophet by the name Jack Wilson—and, during a solar eclipse in 1889, he told Mooney, he fell asleep in the middle of the day and was "taken up to the other world." There he saw the creator and "all the people who had died long ago engaged in their oldtime sports and occupations, all happy and forever young." God sent him back with orders to instruct his people to "be good and love one another, have no quarreling, and live in peace with the whites." And to dance, and thereby hasten their deliverance.

Mooney understood that what he was witnessing was not an obscure outbreak of primitive superstitions but a tide of religious transformation crashing across a continent devastated by conquest. He didn't put it quite this way, but the Ghost Dance was a way of resisting that obliteration, of enlisting the dead to stand alongside the living—and not just to stand, but to sing with them, to dance with them, to embrace them and help them to survive. Mooney knew that this had precedents. "And when the race dies crushed and groaning beneath an alien yoke," he wrote, "how natural is the dream of a redeemer." It was as natural in the foothills of the Sierra Nevadas as it had been in Palestine nearly two thousand years before. Had not Jesus also promised resurrection into eternal life to a people suffering a cruel occupation? Had Christ's message to his followers—to love one another, to turn the other cheek, and to render unto Caesar the things that were his—been all that different? "The doctrines of the Hindu avatars," Mooney concluded, "the Hebrew Messiah, the Christian millennium and the . . . Indian Ghost Dance are essentially the same, and have their origins in a hope and longing common to all humanity."

This was blasphemy. The entire self-understanding of European civilization had been meticulously constructed to avoid such a comparison. The religions of the world could be considered comparatively when arranged on a staircase, a ladder, or a pyramid, but not grossly unranked on a flat and open plain. Powell took the unusual step of disavowing his subordinate's conclusions. In introductory remarks published alongside Mooney's original report, Powell struggled to redraw the proper boundaries between "the superior race" and its delusional, primitive charges. He dismissed the Ghost Dance as a "curious evanescent cult, which seems rather a travesty on religion than an expression of the most exalted concepts within human grasp." More broadly, Powell warned that "caution should be exercised in comparing or contrasting religious movements among civilized people with such fantasies" as the Ghost Dance. Mooney's lyric excesses notwithstanding, no meaningful analogy was possible: "In mode of thought and in coordination between thought and action, red men and white men are separated by a chasm so broad and deep that few representatives of either race are ever able clearly to see its further side."

Especially when it is filled with corpses. In the fall of 1889, the Sioux, like many other tribes, sent delegates to visit Wovoka. They returned to their reservations in South Dakota after the spring thaw, bearing a message that their people greeted with hunger. The Sioux had been, in Mooney's words, "the richest and most prosperous, the proudest, and withal, perhaps, the wildest of all the tribes of the plains." They had fought three fierce wars against the U.S. government in less than thirty years, seen the buffalo on which they relied exterminated and the territories promised to them by treaty shrink to three small, noncontiguous shards. "Then came a swift accumulation of miseries," Mooney wrote: cattle disease, crop failure, one epidemic after another, famine assisted by acts of Congress.

Within a few months of the emissaries' return from Nevada, the Ghost Dance had been embraced by the majority of the tribe. Deliverance would arrive, Wovoka had promised, the following spring. "There was another world coming, just like a cloud," Wovoka had said, according to the Oglala Sioux holy man Black Elk. "It would come in a whirlwind out of the west and would crush everything on this world, which was old and dying. In that other world there was plenty of meat, just like old times; and in that world all the dead Indians were alive, and all the bison that had ever been killed were roaming around again."

In August of 1890, worried that the new faith heralded a general insurrection, the agent charged with supervising the Pine Ridge reservation attempted to stop a Ghost Dance in which two thousand Sioux were participating. "A number of the warriors," Mooney wrote, "leveled their guns toward him and the police, and told him they were ready to defend their religion with their lives." The agent retreated. That November, the first federal troops arrived. In

December, native policemen backed by one hundred U.S. cavalry-men attempted to arrest the Lakota warrior and visionary Sitting Bull at his home on the Standing Rock reservation. In the confrontation that followed, Sitting Bull was killed.*

Fearing a broader repression, a group of several hundred Sioux led by the Miniconjou Lakota chief Spotted Elk, also known as Big Foot, fled their homes on the Cheyenne River and hid out in the Badlands to the south. Starving and freezing, the refugees were returning to the Pine Ridge reservation on December 28 when they came across a detachment of troops from the Seventh Cavalry. Spotted Elk, sick with pneumonia, raised a white flag and agreed to an unconditional surrender. His people set up camp beside a creek called Wounded Knee. The next morning, December 29, Col. James Forsyth, commander of the Seventh Cavalry, ordered his soldiers to search the Sioux camp for guns. This endeavor, in Mooney's diplomatic telling, "created a good deal of excitement among the women and children, as the soldiers found it necessary in the process to overturn the beds and other furniture of the tipis and in some instances drove out the inmates."

One can imagine that the "excitement" was not the fun kind, though it would be hard to guess from Mooney's report. In religious matters, he took pains to seek out native witnesses and informants, but in his account of the events at Wounded Knee, he relied almost

---

* Mooney wrote of the old warrior—who after the Battle of Little Big Horn had been cast in a role similar to the one played by Osama bin Laden in the post-2001 United States—with as much qualified respect as he could permit himself: "He has been mercilessly denounced as a bad man and a liar; but there is no doubt that he was honest in his hatred of the whites, and his breaking of the peace pipe, saying that he 'wanted to fight and wanted to die,' showed that he was no coward. But he represented the past. His influence was incompatible with progress, and his death marks an era in the civilization of the Sioux."

entirely on military and government sources. Perhaps he understood that the issue was too politically dangerous to treat otherwise. Perhaps he knew that ethnological methods could only be applied to Indians. Whites got to speak for themselves. Perhaps, when it came to bullets flying, Mooney simply chose a side.

For all his efforts to portray the soldiers' actions in a sympathetic light, his narrative is nonetheless damning, and all the more unsettling for its many vacillations. "It is said," Mooney wrote, his use of the passive voice an index of his discomfort, that "one of the searchers now attempted to raise the blanket of a warrior." Another young brave, it is said, "drew a rifle from under his blanket and fired at the soldiers," who fired back with their own rifles and let loose "a storm" of 42 mm explosive artillery shells "at the rate of nearly fifty per minute, mowing down everything alive." (The accounts of the few surviving Sioux were quite different, and involved a deaf man who did not understand the soldiers' orders and a gun that went off by accident in the ensuing confrontation. So it is also said.) Twenty-five soldiers and most of the Sioux men were killed in the first barrage. The women and children who were able fled into a dry ravine. Some of them made it as far as two miles before the soldiers caught up with them.

They were found there when the troops returned on New Year's Day, accompanied this time by civilians hired to bury the dead. It had snowed for three days straight, covering the bodies and the blood. Four infants were found alive beside their murdered mothers. One of them survived. The soldiers and the men in their employ dug a long, deep trench. "Into it were thrown all the bodies, piled upon one another like so much cordwood, until the pit was full, when the earth was heaped over them and the funeral was complete. Many of the bodies were stripped by the whites . . . and the frozen bodies were thrown into the trench stiff and naked. They were only dead Indians."

Mooney estimated that three hundred Sioux were killed at Wounded Knee. Twenty of the soldiers who took part in the massacre would be awarded Medals of Honor. Their leader, Colonel Forsyth, would be exonerated of any guilt in the matter and promoted to the rank of major general. Mooney was as careful as he could be. The whole affair, he judged, was "simply a massacre." He stressed that "the first shot was fired by an Indian," which made the Sioux "responsible for the engagement." Nonetheless, Mooney judged, "the wholesale slaughter of women and children was unnecessary and inexcusable." His attempts to mitigate even this compromised judgment are painful now to read: "In justice to a brave regiment," he wrote, "it must be said that a number of the men were new recruits, who had never before been under fire, were not yet imbued with military discipline, and were probably unable in the confusion to distinguish between men and women by their dress."

Perhaps Powell pressured him to write it. Perhaps he gave in to some part of himself that in better moments he could not help but be ashamed of. Perhaps he was clever and calculating and, untroubled by conscience, knew precisely how far he could go. Perhaps he was so uncomfortably split that he did not see the contradictions and believed every word he wrote was equally true. In any case, the massacre effectively ended the Ghost Dance. God had not saved the Sioux. The disaster that came was not the one the prophet promised. "A people's dream," said the Oglala Sioux visionary Black Elk, had "died in bloody snow." The whites were not wiped out, but thrived.

By 1896, when Mooney's nearly five-hundred-page report was published in the second volume of the Bureau of Ethnology's *Fourteenth Annual Report*, several tribes including the Sioux had stopped performing the Ghost Dance. Mooney concluded that they were disappointed by the failure of Wovoka's prophecies. That is almost certainly true of the Sioux, but maybe it was also that the Ghost Dance had accomplished what it could. It had drawn the tribes together and allowed them to find not only devastation but strength in what they had lost. Having done that, they didn't need it anymore. The Cheyenne and Arapaho and the other tribes exiled to the Oklahoma reservations were dancing still, Mooney reported, but without "the feverish expectation of a few years ago." The Paiute were still dancing one year prior to the report's publication, but Mooney didn't know what had happened since. "As for the great messiah himself," he wrote, "when last heard from Wovoka was on exhibition as an attraction at the midwinter fair in San Francisco."

Bison skulls, 1870

☙

It's worth asking again. Is it possible to write without plundering? To leave something behind that is true? I don't mean writing that is innocent. Such a script could have no meaning. I mean writing that takes sides, without compromise or dissemblance or theft, that stands not only with the living but with the dead, with everyone and everything that this society is working to erase.

George Laird danced the Ghost Dance too. It arrived among the Chemehuevi in 1890, which would have made him about eighteen. Both the Chemehuevi and the Mohave hold strong prohibitions against speaking of the recently dead, which may be why the Ghost Dance did not establish deeper roots among them. George apparently said little about it, except, Carobeth wrote, that he "was not seriously impressed by its teaching," and joined the dancing only "for fun."

But aren't the enthusiasms of youth often an embarrassment to the old? Perhaps, after the failure of Wovoka's prophecy, he preferred to downplay the earnestness of his involvement. Even lighthearted acceptance of an apocalyptic creed would likely have been inconceivable to the generations that preceded him. So long as time wheeled smoothly through the seasons, its interruption, destruction, and restoration would hardly have seemed desirable. The trauma of conquest broke all that. By George's adolescence, "with the influx of white settlers and the assertion of governmental authority, the curtailment of hunting ranges, the frightful smallpox epidemics," Carobeth wrote, "time had become suddenly, frighteningly linear."

The Ghost Dance, she judged, "at once expressed and alleviated the sense of impending doom." Perhaps it was something more than that, a way of transmuting death into survival, despair into

togetherness and even hope. Viewed in retrospect, from a distance
of forty years, that might have felt like fun.

☙

I needed to get out of my head this afternoon, and out of the apartment, so I got in the car and drove south across the city toward Sloan
Canyon, where I'd heard there were ancient petroglyphs carved into
the rocks. The GPS directed me through a massive subdivision of
faux-Italian, tile-roofed homes. Inspirada, it was called. I took Via
Firenze to Via Contessa to Democracy Drive. None of it looked
more than a few weeks old. The roads were freshly paved, the curbs
and sidewalks not even poured, but there were box stores up already,
built for neighborhoods that didn't yet exist.

For miles around, to the very edge of the mountains, the desert
was being devoured to make way for new construction. Bulldozers
and earthmovers had torn up every shred of life, leaving only the dry
dirt, ridged from the steel treads of the machines. I saw them parked
in long lines like soldiers at rest: bright yellow graders, excavators,
backhoes, scrapers arrayed behind a water tank painted red, white,
and blue. PATRIOT CONTRACTORS, the side of it said. The machines were all idle, but sprinklers were nonetheless spraying water
in high arcs over the ruined earth. To keep the dust down, I guess.

I got lost. The streetscape was evolving too quickly for the GPS
to make sense of it. I drove down a long dirt road that led to a water
tower surrounded by a wall topped with barbed wire. I got out of
the car, walked a bit, tore off a sprig of creosote, and sniffed it. There
was no one around to ask, only cameras and machines. I stopped at
a fire station. It was as new as everything else. Someone had reckoned on the certainty of flames. The fireman who answered the door
was friendly. I tried not to smile when he advised me to get back on

Democracy and follow it all the way around the bend, past where it narrows, and to be sure to turn off before it ends.

Finally I found the trailhead. I had less than an hour of sun left, so I hurried up the wash. The plants were familiar from Joshua Tree: cat's-claw and yucca, desert indigo and Mormon tea. The wash narrowed as it climbed. Walls of rough volcanic rock rose along both sides. Maybe two miles in, I spotted the first petroglyphs: two bighorn sheep in profile, carved into an outcropping high above the floor of the wash. A few feet farther they were everywhere I looked. More bighorns, lizards, branches, insects, stick-figure humans with long fingers raying out like stars. Some carried bows and arrows. Most, though, were abstract and indecipherable: wavy lines and circles or spirals, figure eights and shapes I could not make sense of. Some were many thousands of years old, chiseled into stone by people who lived here long before the Paiute.

I hiked back out, slowly, the sky going pink above the canyon walls. After the first mile I could see Las Vegas laid out in the valley beneath me. Again, that gleaming net of gold. It was nearly dark when I got to the car. I didn't want to go home, didn't want to drive back into the city, so I sat there as long as I could and watched the moon rise over the hills while the stars found their places in the sky.

❧

Another school shooting. Seventeen dead in Parkland, Florida, all but three of them children. And another new study found that even if greenhouse gases are immediately reduced, sea levels will continue to rise for the next three hundred years.

It was eighty degrees in Washington, D.C., today. It's February.

❧

This is how they found Spotted Elk, also known as Big Foot, chief of the Miniconjou Lakota, on New Year's Day of 1891. He was sick with pneumonia at the time of the massacre. I have found accounts saying he was shot while stepping out of his tent, others that claim he was killed before he could rise to his feet. His posture in the photo is hard to make sense of. Perhaps the snow had melted from beneath him. Perhaps it was there still, but rendered invisible by the lack of contrast in the image. Perhaps he was leaning forward even in death, attempting to see what lay before him.

5.

The first explicit articulation of the modern notion of progress is generally agreed to have appeared in a speech delivered in 1750 by the brilliant political economist Anne Robert Jacques Turgot, then just twenty-three. It is surely no coincidence that an early evangelist of economic liberty—"All branches of commerce ought to be free, equally free, and entirely free," he wrote in 1773—would also be the first to lay out the ideology that would everywhere accompany the spread of capitalism, this mystic doctrine that envisions history traveling on a one-way path toward some as yet unimaginable perfection.

Turgot, who would go on to serve as finance minister under Louis XVI, would have an opportunity to test his economic theories if not his conviction that mankind was destined to shed its every flaw. In 1774, acting as the controller general of finances, he issued an edict abolishing all restrictions regulating the trade of grain in France. The next year's harvest was poor. Merchants, freed from laws that had proscribed stockpiling, hoarded wheat to drive up prices. Famine ensued. Riots broke out. In what would in retrospect be seen as a prelude to the French Revolution, crowds, many of them led by women, forced landowners and merchants to sell their grain at prices they deemed reasonable. The Flour War, as it would be known, ended with a fierce repression. The government called in twenty-five thousand troops. Within a year, court intrigues would lead to Turgot's dismissal. His reforms were soon reversed. Perhaps he was lucky. Turgot died a private man, of gout, twelve years before the monarch met the guillotine.

At the very beginning of the lecture in question, "A Philosophical Review of the Successive Advances of the Human Mind," Turgot made a surprising move. He had no sooner distinguished the sort of time that governs humanity from that which governs nature—the latter following the timeless and cyclical rule of death and regeneration while human events succeed one another in a linear and "ever-changing spectacle" as the species lurches from "infancy" to "greater perfection"—than he evoked the specter of "the Americans." This comes on the second page. He was not thinking of the well-armed citizens of the current American polity, but of the original inhabitants of the hemisphere, whose presence posed something of a problem. If history was to be narrated as the story of the growing perfection of mankind, how to explain all those who seemed to lag behind, the tribes of so-called primitives scattered across the jungles, plains, and desert wastes? (Adam Smith, Turgot's contemporary and fellow preacher of economic liberty, made an analogous move in the earliest paragraphs of *The Wealth of Nations*, in which he evoked "the savage nations of hunters and fishers," accusing them of infanticide and comparing them unfavorably with "civilised and thriving nations.")

We might think that progress is a theory of history, and hence of time, but very nearly the first thing Turgot did was transpose time onto space by summoning the Americas as the place of the past. He explained the apparently uneven development of humankind and the anachronistic existence of the people he regarded as savages as a result of naturally occurring inequalities: "Nature, distributing her gifts unequally, has given to certain minds an abundance of talent which she has refused to others." Differing environmental circumstances allowed those original talents to develop at different rates, "and it is from the infinite variety of circumstances that there springs the inequality in the progress of nations."

Before it was anything else, the doctrine of progress was a theory of white supremacy. Better put, since strictly racial theories would not arise for another few decades, it was a cocksure expression of what even then was a highly parochial and amnesiac variety of chauvinism, a way of celebrating European dominance by anchoring it in time, and rendering Europe, and specifically Bourbon France, the very apotheosis of human achievement. In a parallel move three-quarters of a century later, Hegel would claim that same honor for the Prussian monarchy. One and a half centuries after that, just as the carbon dioxide in the atmosphere was sliding past 350 ppm, such lights as Francis Fukuyama would do the same for capitalism under American liberal democracy. As an ideology of overconfident elites, progress would prove itself remarkably resilient.

It's worth recalling that the first Europeans to lay eyes on the great cities of the Americas did not see them as Turgot did. Francisco Pizarro, the conqueror of Peru, wrote to King Charles V that the Inca capital of Cuzco, which Pizarro plundered and largely destroyed, "is so beautiful and has such fine buildings that it would be remarkable even in Spain." Cortés apologized to the same monarch that he did not have the literary skills to adequately describe the marvels of Tenochtitlán, with its temples and wide causeways rising from the waters of Lake Texcoco, its fragrant gardens, great public squares, markets filled with riches: "I am fully aware that this account will appear so wonderful as to be deemed scarcely worthy of credit; since even when we who have seen these things with our own eyes are yet so amazed as to be unable to comprehend their reality."

Cortés's soldiers, his lieutenant Bernal Díaz del Castillo wrote, "had been in many parts of the world, in Constantinople, and all over Italy, and in Rome," but they had never seen anything so wondrous as the Aztec city. "I stood looking at it, and thought that no land like it would ever be discovered in the whole world," Díaz wrote. "But today all that I then saw is overthrown and destroyed; nothing is left standing."

Just last week I read in the papers about the discovery of the remains of villages, roads, and fortified settlements long concealed by foliage in the Amazon, traces left by as many as a million people who had all disappeared at about the same time—roughly when Europeans arrived on the continent. "Diseases travel much faster than people," the lead archaeologist told *The Guardian*. The Amazon's inhabitants may have been wiped out before the Portuguese ever reached the area.

It's a good trick, really, so seamlessly clever that Turgot and many millions since didn't see it as a trick at all: allow biological agents to help you conquer half the world, slaughter and enslave whomever the microbes let live, declare the hobbled remnants of the civilizations you have destroyed *primitive* and *savage* and interpret the degraded manner in which many of your victims are forced to survive as evidence of your inherent superiority, your right to rule over them and to continue to exploit them under cover of *civilizing* motives. Progress, it's called.

Turgot went on to sketch out a narrative that should by now be quite familiar: the mantle of civilization passed from the once-great civilizations of Egypt, India, and China, all strangled by their own despotism, and onward, via the Phoenicians—in themselves mere "agents of exchanges between peoples"—to Greece and then Rome, until the latter empire, a victim of its own descent into tyranny, "at last suddenly collapses" under the attacks of opportunistic hordes.

The rise of Islam merited brief condemnation, almost parenthetically, as "a raging torrent which ravages the whole territory from the Indian frontiers to the Atlantic Ocean and the Pyrenees." The glories of Baghdad and al-Andalus were passed over without comment, though Turgot acknowledged that Islamic scholars did Europe the service of transmitting Europe's past to its future by disseminating "the feeble sparks" of Greek wisdom that they had managed to preserve. That was all it took. "The treasures of antiquity, rescued from the dust . . . summon[ed] genius from the depths of its retreats. The time has come," Turgot enthused. "Issue forth, Europe, from the darkness which covered thee!"

Messianic thinking, that child of disaster and offspring of the oppressed, takes on here, in the hands of the triumphant, a strange, tumescent form. Expectation is still its guiding passion. Paradise awaits, but it now belongs to human—read European—reason, unaided by the divine. Knowledge builds upon its own accomplishment. Scientific and technological discoveries pile one atop the other. "The scaffolding rises with the building," Turgot wrote. Time is not a scar, but a boast, and a promise. "Time," urged Turgot, rhapsodic, "spread your swift wings!"

His discourse concludes with a paean to the king, not the one Turgot would later serve and who would still later lose his head in the Place de la Concorde, but his predecessor, also named Louis, who died in his own bed in the palace at Versailles, of smallpox—the same virus that helped reduce the population of the Americas by as much as 95 percent. "O Louis, what majesty surrounds thee!" Turgot crooned. "Century of Louis the Great, may your light beautify the precious reign of his successor! May it last for ever, may it extend over the world!"

☙

All those deaths—the 95 percent of the hemispheric population that did not survive the earliest encounters with the "greater perfection" of Europe—may have actually altered the climate. Other scientists blame sunspots or volcanic eruptions, but so goes one of the current theories explaining the advent of the so-called Little Ice Age, the brief episode of global cooling that peaked between about 1550 and 1750. The rapid depopulation of the Americas meant that large-scale agriculture all but ceased throughout the hemisphere. No one was left to burn woods and grasslands to clear stumps and brush for farming, or at least not on the scale that they had been a few years earlier. The land recovers swiftly without us: More than a hundred million acres of once-cultivated land grew back as forest and savanna—the Edenic, apparently virgin woods and plains that European settlers found spread out before them like a dream. Which meant billions of new trees, grasses, vines, and shrubs sucking carbon dioxide from the atmosphere: by one estimate, the post-conquest reforestation of the Americas sequestered between five and ten trillion metric tons of carbon, causing global CO_2 levels to fall by 2 percent and temperatures in the Northern Hemisphere to tumble.

Beginning in about 1570, Europe endured a series of deadly winters followed by cool summers in which the sun barely shined. Grain harvests failed. Between 1630 and 1710, France suffered five dreadful famines, one of them stretching over three years. The feudal system, which depended on the production of an annual surplus of grain that peasants would pay as a tributary tax to landed nobles, began to unravel. So did the ideologies it had spawned, and on which it rested. By one account, 1,265 food riots tore at France between 1661 and 1789. There was one in 1752, two years after Turgot rhapsodized his way through his "Philosophical Review," just as the Little Ice Age was lifting and the climate, for a little while anyway, was stabilizing once more.

It's windy today. Two plastic bottles, an empty snack-sized bag of Doritos, and a paper burger wrapper have joined the dog shit and the yellowed bamboo leaves on my backyard square of Astroturf, gifts blown in from the lots and the alleys. I left town for two days and came back to find the weather warmer. Before I left it had been dropping almost to freezing each night. Everyone I talked to was complaining. In Europe it had been colder still. Snow fell as far south as Rome. In England, temperatures hit record-breaking lows. The British army had been called out to ferry health workers over frozen roads. All of this, apparently, because melting sea ice and lack of snow cover in Siberia had caused a high-pressure front to drift north. The polar vortex, which normally traps cold air over the Arctic, had collapsed, allowing the low pressure and cold air to drop south over Europe and the American West. Temperatures in the Arctic, meanwhile, are fifty degrees higher than they usually are. It's still winter, and it's above freezing at the North Pole.

It is now such a reflex to consider past civilizations stupid and superstitious—or stupider, at least, than we are—that it is hard to imagine the relief that men of Turgot's era must have felt on unburdening themselves of the past, launching themselves into a limitless future. Previous generations of human beings in nearly every corner of the planet had regarded their ancestors with veneration, and what a drag it must have been to have to lug those old farts and all their dumb ideas everywhere, and always with a smile. Turgot's ecstasy broke through in his punctuation, and in the impatience of his syntax: "What ridiculous opinions marked our first steps! How absurd were the causes which our fathers thought up to make sense of what they saw! What sad monuments they are to the weakness of the human mind!"

The first thing he hastened to toss over was the notion that all things are alive and infused with divinity. This, by his reckoning, was one of "those delusive analogies to which the first men in their immaturity abandoned themselves with so little thought." Like children, Turgot explained, they imagined that all the things they perceived that "were independent of their own actions were produced by beings similar to them, but invisible and more powerful." Thus, they supposed, superstitiously, "all objects of nature had their gods." Against this, he allied himself, if only half-heartedly, to a sterile monotheism, nodding to the Christian deity later in the lecture, but only in passing, and without a shred of the enthusiasm that he brought to subjects such as Reason, Europe, and France. But the task of denuding the natural world of agency and divinity was apparently an important one, and could not be neglected. For the grand procession of progress to march, the stage had to first be cleared of rivals. All the world must be dead, and man alone alive, rushing to the glory of his fate.

≋

On the morning of July 6, 1962, the United States Atomic Energy Commission detonated a 104-kiloton thermonuclear bomb 635 feet beneath the ground at Yucca Flat, Nevada, about one hundred miles northwest of the Las Vegas Strip. The test, one of hundreds conducted there, was code-named Sedan. It was part of Operation Plowshare, a project meant to explore, in the words of its architect, the physicist Edward Teller, the possibilities for "constructive, peaceful uses of nuclear explosives," namely moving massive amounts of earth more quickly and cheaply than conventional methods would allow. "We needn't take a coastline as it happens to be," enthused Teller at the time. The AEC drew up plans to excavate new harbors for oil tankers in Alaska, to blast a canal across the width of Nicaragua, and to use twenty-three bombs to slice a trench through the Mojave's Bristol Mountains so the railroads and Interstate 40 wouldn't have to loop around them. Nuclear explosions might even one day be used to influence the weather, Teller wrote in an article published two years before the Sedan test. "We're Going to Work Miracles," it was titled. It is a dizzying creed, progress.

Three seconds after the Sedan detonation, a dome of loose earth more than six hundred feet in diameter rose three hundred feet in the air. A bright light flashed across the sky. The cloud kept climbing, rising to twelve thousand feet above the Nevada desert and splitting into two enormous plumes of radioactive dust and ash that drifted north and east, dropping fallout as they went, until they finally passed over the Atlantic Ocean somewhere between North Carolina and Delaware. Some of the highest concentrations, as luck or fate would have it, landed in Washabaugh County, South Dakota, which at the time was almost entirely within the boundaries of the Rosebud and Pine Ridge Indian Reservations—home to, among others, the Miniconjou and Oglala Lakota Sioux.

≈

Even the desert can start to feel small. Edward Teller was born in Budapest under the name Ede, which he changed before emigrating to the United States in 1935, just a few years after his cousin György Dobó changed his name to George Devereux and began doing fieldwork with the Mohave, along the Colorado River, about 275 miles to the south of the desert testing grounds that Teller would so assiduously bomb.

≈

I drove yesterday to the Valley of Fire, a state park about an hour northeast of the city. It was a long week, and a strange one. The Rhino proposed arming schoolteachers. This is how policy seems to work these days: an idea so outlandishly stupid that it is initially floated in the press only as a laugh line—Rambo teachers, a literal wall on the Mexican border—gains traction on right-wing media, crosses over to mainstream outlets, and is eventually accepted as, if not an inevitability, an unmovable part of the discourse. But by the end of the week enough momentum for gun control was building— thanks to the courage and outrage of the teenaged Parkland survivors—that even the Rhino was considering token measures. Wayne LaPierre, the reptilian chief of the National Rifle Association, announced that gun ownership is a right "granted by God." A church in Pennsylvania—the World Peace and Unification Sanctuary, it is called—asked parishioners to bring their assault rifles to services so that they could be blessed, like puppies and hamsters on the feast of St. Francis. Photographs later emerged of the rite. The worshippers wore brass crowns of bullets. In Syria too, shells circled civilians' heads: the Assad regime was bombing East Ghouta, just outside Damascus, killing hundreds. And the Rhino, in a news conference with the Australian prime minister, gloated over a new set of sanctions against North Korea. If that didn't do the trick, he warned, "we'll have to go to Phase Two. Phase Two may be a very rough thing. Maybe very, very unfortunate for the world."

I got off the interstate at the edge of the Moapa Paiute reservation and drove another fifteen miles east into the park. The rocks were a craggy, lumpy sandstone, a shocking red in color, as if they'd been dripped onto the mottled floor of the desert from some enormous paintbrush. I found a trail, hiked in a few miles, ate lunch with my feet dangling over the edge of a boulder, and napped beneath a steep escarpment of red stone. The silence did me good. And the folds and hollows in the rock, the wind rushing through the stern, determined chaos of the landscape. Two ravens flew by, cawing to each other in flight. An owl hooted somewhere out of sight. I walked not fifty feet past a big-horn sheep munching at the shrubs along the edge of a wash. We stared at each other for a minute or two before he went back to his meal.

I made it to the car a few minutes before sunset, when the park closed, but I had time to check out the petroglyphs just off the main road. They were carved into a flat, vertical plane of rock perhaps sixty-five feet up from the desert floor. Park authorities had built a metal staircase so that visitors could see them from the comfort of a viewing platform that was likely designed to discourage vandalism but that had not stopped Johnny, Doreen, and Hermann, among others, from scratching their names alongside the ancient glyphs. They were better preserved than the ones I had seen in Sloan Canyon, and seemed more purposefully arrayed: circles within circles, a cross, bighorn sheep, what looked like drooping Joshua trees, suns, and squiggly lines. The usual official signage explained that they once meant something but we don't know what it was, and please don't destroy them.

I had been thinking about writing all afternoon, what it is for, and whom, wondering why—if I find it hard to believe it will make a difference to this sinking world—I still do it, scratching away at the keyboard, as if someone, one day, would weld a staircase and viewing platform beneath my screen and carve their own names beside these by-then indecipherable runes. I can't say I came up with much. (From George MacDonald's *Lilith*: "I do not know about *world*. What is it? What more but a word in your beautiful, big mouth?") It was dark when I headed down the long, straight road to the interstate again. I kept seeing bursts of light in the distance ahead of me, six or eight miles off, sudden blossoms of white or red or green that hung in the air for a few seconds before falling and fading to black. Fireworks. I was too far off to hear them. I wondered if there was a party in town, a quinceañera or a wedding, but when I got closer I remembered that there was no town, and nothing for miles, just a brightly lit truck stop at the southern boundary of the Moapa reservation. I pulled into the lot beside the gas station and parked among the semis idling there. I didn't see a party, or anything, really, just a few kids in a wide, bulldozed field of dirt, shooting fireworks high into the sky and, occasionally, at each other.

❧

The other great early exponent of progress was Marie Jean Antoine Nicolas de Caritat, better known as the Marquis de Condorcet. A friend of Turgot, and later his biographer, Condorcet became an ardent republican and after the Revolution served as secretary of the Legislative Assembly until that body was dissolved, and the king dethroned, in 1792. In those days, surely, the world must have felt like it was ending, or beginning again. Condorcet welcomed its rebirth. He was at odds, though, with the Revolution's most radical factions, and was sharply critical of the constitution drafted in June of 1793 by Robespierre and Saint-Just. In July a warrant was issued for his arrest.

Condorcet would spend the next nine months in hiding, cooped up in a house on a small street just north of the Luxembourg Gardens. Three months into his confinement, he would be condemned to death in absentia. The Terror had begun. Despite these pressures, Condorcet continued to work on the text for which he would be best known, a book-length treatise of astonishing optimism entitled *Outlines of an Historical View of the Progress of the Human Mind*. "The perfectibility of man is absolutely indefinite," he wrote in the introduction, and "the progress of this perfectibility ... has no other limit than the duration of the globe upon which nature has placed us."

He would not live to see the work published. Early in April of 1794, informed that his capture was imminent, Condorcet fled Paris. Two days later, bleeding from a wound on his leg, dirty and bedraggled from sleeping in the open, he walked into a wine shop in a village just outside the city. He ordered an omelet. When the proprietor asked how many eggs he wanted, Condorcet, who had presumably not eaten for days, aroused the suspicion of his host by answering, "A dozen": only an aristocrat would presume to order an omelet of a dozen eggs. He was dragged to the prison in the commune of Bourg-la-Reine, which at the time was known as Bourg-l'Égalité, or, roughly, Equality-ville. Accounts vary, but either the following morning or the one after that, the Marquis de Condorcet was found dead on the floor of his cell.

While still in hiding in Paris, at work on the *Outlines*, confined to the rooms of the small house on Rue Servandoni, his country at war against most of Europe, hearing frequent news of the beheadings of friends and colleagues, Condorcet surely must have suspected that his own death, and not a pleasant one, was likely near. He nonetheless remained convinced, with a fervor that can only be described as ecstatic, that there was no other path for humankind than that of ever-increasing perfection. The final chapter of the text was devoted to forecasting the outlines of that bright future. Everyone on the planet, he predicted, would "one day arrive at the state of civilization attained by those people who are most enlightened, most free, . . . as the French, for instance, and the Anglo-Americans." This would mean an end to "the slavery of countries subjected to kings, the barbarity of African tribes, and the ignorance of savages."

Condorcet was a brave man, and every bit the liberal hero: he had only the most passionate and eloquent words of condemnation for slavery, the oppression of women, and the brutal exploitation of colonized peoples, but his certitude rested on a deep and unquestioned conviction in the moral superiority of Europe, despite all the dizzying evidence to the contrary. Slavery and the various cruelties that Europe had visited on the people of Africa and Asia would soon surely end, he believed, "and we shall become to them instruments of benefit, and the generous champions of their redemption from bondage." It's quite a fantasy: the dog, having been whipped to a degree that even the master must acknowledge as excessive, will, with a few encouraging pats on the hindquarters, lick his master's magnanimous hands.

Progress, once again, was before all else a doctrine of supremacy, a fresh new faith for a rising era of unchallenged European dominance. It extended through time the inequities of power that had only in the previous two centuries taken shape across the globe, spreading the same cruel hierarchy across all of time and space. It was Europe that held exclusive title to reason and to peace—this despite the conflicts that had wracked the continent over the preceding hundred years: a seven-year war that had killed nearly a million just three decades earlier, the wars over the Austrian and Spanish successions that between them killed another million and a half, and the war that the Revolution itself had sparked, which by 1815 would have taken another five million lives. Nonetheless it was Europe that had lit the way and Europe that would cut through all remaining darkness, Europe that would lead the scattered people of the earth to a future without war or oppression or discrimination based on sex or tribal prejudice. Sanitation would improve, and nutrition, and medicine. Human lifespans would grow longer. One universal language would spread across the earth "and render error almost impossible." Except for the language bit—English functions well enough—this is still the faith that fuels contemporary neoliberalism and the logic of "development": The global extension of European ways, meaning open markets and liberal institutions, will spur technological improvements and improve the lot of all, indefinitely. One day the natives will thank us. It is a powerful and alluring vision. For some, at least. Little wonder that Condorcet clung to it even as the world he knew was being washed away with blood.

Condorcet took one notable step that his friend Turgot did not. In his introduction, he devotes more praise to one advance than any other: the invention of *writing*. Specifically, of alphabetical writing.

After quickly sketching the leap from hunter-gatherer societies to agriculture and the birth of property, he conjectured, as Turgot had before him, that the resulting social inequalities produced a class that was not compelled by necessity to work, and was hence at liberty to observe, experiment, and invent. (In their abundant leisure, the rich have rarely found time to notice the inventiveness of those whose labor makes their comforts possible.) As their societies grew more complex, the members of this class "felt the necessity of having a mode of communicating their ideas to the absent." In response to this pressing need, some brilliant soul or souls began to *write*.

It was at first a crude affair, pictorial symbols that represented the objects for which they stood by a process akin to metaphor. For Condorcet, the real and decisive innovation, the one invented by "men of genius, the eternal benefactors of the human race," was alphabetic writing. He did not likely know of Maya glyphs or Mojave petroglyphs and would not have thought much of them, but this distinction allowed him to relegate the Chinese and the Egyptians to the backward realm of metaphor, and to elevate the Greeks and all those who followed them into the paradise of phonetic abstraction. Such thinking was not new: as early as the sixteenth century, the Jesuit José de Acosta, who had spent seventeen years in Mexico and Peru and was familiar with both the Maya codices and the *quipu* of the Inca, made a distinction between written signs that referred to words and pictorial signs that referred directly to things. The latter, he wrote, "are not called, nor are they in reality true letters, even though they are written, just as one cannot say that a painted image of the sun is writing." It goes without saying, Acosta continued, that "no nation of Indians that has been discovered in our times uses either letters or writing."

Astronomy was nice and agriculture convenient, but no other human accomplishment mattered more. Writing was the thing. It was the key to the map sketched out by the long, forward march of progress. "It is between this degree of civilization and that in which we still find the savage tribes," wrote Condorcet, "that we must place every people whose history has been handed down to us … an unbroken chain of connection between the earliest periods of history and the age in which we live, between the first people known to us, and the present nations of Europe."

ℤ

Ha! Kim Jong-un invited the Rhino to Pyongyang for talks and the Rhino said yes.

ℤ

Condorcet was surely not aware of it, but it was in April 1784, ten years to the month before his death, that James Watt patented the double-acting steam engine that would within a few decades spread throughout Great Britain, over the Channel to the Netherlands, France, and Russia, and across the Atlantic to the United States. By the close of the next century, the world would run on coal. Human beings, frail and featherless, had proven ourselves something more than clever predators. We could make the skies darken at noon, and brighten the night, and shift the very winds and the currents of the seas. Who else but man could melt a glacier? Who else but man could burn it all? Condorcet was right: Europe indeed had lit the way.

ℤ

I went away for the weekend to visit family. Leaving Vegas makes me happy and seeing them makes me happier still, but it means that I don't write. If I let too much time pass without writing, I start to fray. I get depressed, a bit crazy. It all piles up so quickly, and this is the only way I know how to sort through it, to cut it down. So though I miss them, I am happy to be here. Here meaning this place, on this page, in these words. Not Las Vegas, where I also am.

I came across a rare bit of local public art this morning, an equestrian monument cast in bronze. It stood outside the parking lot of a community center in a district of low warehouses behind razor-wire fencing. Letters carved into a stone in front of the statue identified it as a gift of the Rio Hotel & Casino. The rider was Rafael Rivera, "discoverer of the Las Vegas valley," a distinction the Paiute and Chemehuevi would likely dispute. Rivera was a scout for a trading party headed for Los Angeles in 1829. This was still Mexico then. Rivera was sent off to find water, and he did. He wandered for more than a week before he found this valley. The springs had not run dry yet. It was green here then, lush enough that they called it "the meadows," which is the name it still bears. The sculptor gave Rivera Anglo features and a dull, gormless expression, as if he could not understand what he was seeing.

Early in the morning, it's easier to remember that this is desert. The doves and pigeons are just waking. I try to concentrate on them as I run, to focus on the way pigeons go motionless as they approach the ground, their tail feathers spread, wings behind them in a taut, frozen *V*. I ran east toward the sun rising behind the mountains, the sky still pinker than it was blue. I tried to squint away the buildings and the streets. It didn't work. This morning the Rhino fired his secretary of state, whose fate has been sealed since October, when he called the Rhino a "fucking moron." The court dramas have been nonstop all week, news breaking every day about fresh departures from the White House and the latest from Stormy Daniels, the porn star with whom the Rhino had a brief affair in 2006. Vladimir Putin, in an interview with NBC news, denied that Russia had interfered with the 2016 U.S. election and suggested that perhaps it was the Ukrainians, or the Jews. The Rhino has not voiced a word of complaint. He'll replace the outgoing secretary of state with the current CIA chief, who is much keener on punishing Iran, and less given to contradicting his boss. The CIA chief will be replaced with his deputy, who directly oversaw the torture of CIA prisoners at a secret site in Thailand. Rumor has it that the Rhino plans to replace his national security advisor, who is regarded by D.C. insiders as one of the few brakes on the Rhino's most recklessly bellicose moods, with a famously abrasive and immoderately mustachioed neoconservative who has made a career of reckless bellicosity.

I ran home on Sunrise Avenue, past block after block of stained stucco apartments, about a third of them boarded up, street numbers painted with black stencils on the walls. I turned up Fremont Street, named for the first Anglo-American to pass through the Las Vegas valley. Fremont is the main drag downtown, home now to many fine abandoned motels. There are counties in four states named for him too, and at least a dozen towns, three mountains, and a river. Rafael Rivera got the park behind the community center, plus the service road that runs alongside the 215 freeway.

(To be fair, being white and stumbling through a valley that already bore the name "Las Vegas" was not John C. Frémont's only claim to fame. In April 1846, his band slaughtered somewhere between 120 and 1,000—they did not count, and the recollections of Frémont's men varied widely—men, women, and children, probably of the Wintu tribe, on the banks of the Sacramento River near what is now Redding, California. Ten years later he would become the Republican Party's first candidate for president.)

I scared up a gathering of mockingbirds in the corner of an empty lot. David Byrne sang through my earbuds, "Time isn't holding up / Time isn't after us." A few blocks down, I was surprised to spot an osprey flapping slowly above me in the sky, a fish hawk far from any body of water larger than the fountains in front of the casinos, far even from Lake Mead and the detergent-scented eddies of treated wastewater that flow through the Las Vegas Wash. The pigeons scattered beneath it, panicking as the larger bird flew past.

≈

In October 1793, the same month that Condorcet was condemned to death, France's National Convention, which had replaced the Legislative Assembly one year before, voted to adopt a new calendar. I've seen a copy of it, or a photo of one, the paper brown with age, frayed and dog-eared, the words "Calendrier Républicain" printed in black ink on the cover, and beneath them, in smaller type, and in French, "Decreed by the National Convention, for the IInd year of the French Republic." The year before, the Convention had begun counting the years anew, dating them to the birth of the Republic rather than the birth of Christ. Time had zeroed out and started again.

How giddy they must have felt, how brave and how light. Everything was new, or would be soon. The cobwebs of the old regime and the centuries of priestly superstition on which it rested would all be swept away. The Republican calendar, designed by a committee of politicians, scientists, mathematicians, and a writer, would be clean and exact, suitable to the age of reason that the Revolution had ushered in. Its predecessor, one committee member (the writer) complained, "exhibited neither utility nor method; it was a collection of lies, of deceit or of charlatanism." The new one would do away with all that, scrapping the jumbled saints' days and festivities and all the religious holdovers of the Gregorian calendar, stripping time of the accumulated prejudices of antiquity. It's a curious notion, really, that time should not be tainted by the past.

The Republican calendar would divide the year into twelve months, each one composed of three ten-day weeks, plus a week of either five or six "complementary days" to keep the calendar in synch with the sun. Acknowledging the twelve yearly orbits of the moon around the earth would be the committee's only accommodation to the solar system's disregard for decimals: each day would be subdivided into ten hours, every hour into one hundred minutes, the minutes into one hundred seconds each. The pagan origins of the names of the months—*janvier* for the Roman god Janus, *février* for the ancient Roman purification festival *fevrua*, *mars* for the god of war—would be replaced with neologisms drawn from French, Latin, and Greek describing climactic conditions around Paris at the appropriate time of year, i.e., Brumaire, or "misty" in the fall; Nivôse, or "snowy" in midwinter; Floréal to coincide with spring's first blooms. The days of the week would be similarly cleansed in favor of austere, ordinal numbers: *primidi, duodi, tridi,* etc. Realizing that the peasantry might miss the festive wheel of saints' days—and the revelries that went with them—the committee devised a separate rural calendar, naming every day of the year after "useful products of the soil, the tools that we use to cultivate it, and the domesticated animals, our faithful servants in these works." In this abundance of whimsy, 2 Frimaire was named for turnips, 8 Nivôse for manure, 15 Vendémiaire for donkeys, and 10 Thermidor for the watering can.

For all its semi-comic joylessness and kitsch, it might be easy to miss that this rationalization also functioned as an ethnic cleansing of traditional timekeeping, which for centuries had acted as a depository of disparate and forgotten forms. The days might be named for Roman gods, but there were seven of them because the Babylonians deemed that number significant, presumably because seven easily visible heavenly bodies circle along the ecliptic: the sun, the moon, Mars, Mercury, Jupiter, Venus, and Saturn. The days are divided as they are because the Egyptians used a duodecimal arithmetic, splitting the night into twelve roughly equal periods, each marked by the rising of a different star. We divvy hours and minutes into sixty pieces each because the Babylonians and Sumerians employed a sexagesimal mathematics. There is nothing more rational about the number ten except that the Greeks and Romans based their arithmetic upon it, and they were the elected paragons of reason. It is difficult to imagine any logical advantage for the use of Roman rather than Arabic numerals to mark the years, as the designers of the new calendar preferred, except to retroactively cement this alliance with the Empire of Reason.

I am not suggesting that this was an intentional or even conscious goal on the part of the calendar's designers. The association of parochially European habits with the universality of reason did not have to be conscious to be effective. It has historically been so effective precisely because it remains an unconscious constituent of an entire system of thought. And of power. Fabre d'Églantine, the committee's resident poet and playwright, who was largely responsible for naming the days of the rural calendar, did not likely pause to consider that there is nothing universal about the weather in Paris. He had only to open a window or walk in a park. But even if you call the month Frimaire, it will not be frosty in Cayenne in December, and horse chestnuts, honored on 23 Germinal, will not thrive in any month in Martinique.

In the end the calendar died with the Republic. Napoleon officially tossed it aside one year after his coronation. The days would march past as they had before the Revolution. Or at least with the same old names. From almost the beginning, though, it had been evident that the new calendar did not function very well. It is safe to say that no one who lived off their labor preferred a ten-day wait for the weekend over the old sabbath every seventh day, and that the peasantry found 8 Nivôse, the Day of Manure, and 30 Prairial, the Day of the Hand Cart, to be insulting and inadequate surrogates for the days of rest and feasting that they replaced.

Even on a theoretical level, the calendar was flawed. The committee had introduced a contradiction into its design that rendered it more confusing, and less functional, than many calendars devised by "primitive" astronomers. The years, according to the committee's decree, would start on the autumn equinox. A leap day would be added as a sixth "complementary day" every fourth year. But the autumnal equinox does not reliably coincide with the first day of the year every four years. To remedy this, they could either embrace nature or abstract reason, but not both. They could pin the calendar on an astronomical event, the equinox, and have the leap year vary, or they could impose an artificial regularity and abandon the calendar's anchor to the rhythms of the heavens.

Charles-Gilbert Romme, the head of the committee, suggested the latter tactic, borrowing the Gregorian calendar's method of adding an additional day to years divisible by four (unless they happened to be divisible by one hundred but not by four hundred). His solution was inelegant, and tainted by association, but it did have the virtue of working. It wouldn't matter. Romme's proposal was never adopted. He was arrested in the spring of 1795, year III. That June 17, or 29 Prairial, he stabbed himself to death before he could be led to the guillotine, which had taken the head of his colleague, the poet and playwright Fabre d'Églantine, the previous 16 Germinal. It was Lettuce Day, by his own pen.

I read in the paper this morning, or on my phone, really, that rising sea levels are threatening the mysterious sculpted heads on Easter Island. Already the beaches there are washing away. The waves are revealing old tombs, exposing ancient bones. The cliffs are eroding. If they go, the giant heads, remnants of a civilization that has long since collapsed, will go too. The article ended on a note that I couldn't help but find hopeful. The people of Easter Island were doing what they could to get ready, collecting data, making plans. "We have been here 1,000 years," a local architect named Hetereki Huke told the reporter. "Believe me," he said, they had been through times like this before, and they knew: "The world isn't ending."

$$\approx$$

Civilizations come and go. Who but historians and archaeologists think about the glories of the Nanda or the Wari or the sixteen-hundred-year rule of the Chola? Most, though, do not simply vanish. People fight to stay alive, cultures too. When the Classic era of Maya civilization collapsed in about A.D. 900, its cities were abandoned, but the Maya did not disappear. They just moved, shifting to the peripheries of the old Classic era terrain: to the northern edge of the Yucatán Peninsula and south into the Guatemalan highlands. For centuries after the Spanish conquest, they continued to refuse to disappear. In 1546 the Maya of the Yucatán revolted, slaughtering their conquerors in large numbers before being slaughtered themselves. Three hundred years later they nearly drove whites and mestizos from the peninsula and created an independent Maya state that would survive until the early twentieth century. In Guatemala, Maya in the Petén region managed to keep the Spanish at bay until the late 1600s. The K'iche' briefly won independence in an 1820 revolt. The Mexican state of Chiapas saw major indigenous rebellions in the eighteenth, nineteenth, and late twentieth centuries. I was there for the tail end of one of them.

It was late August 1995. On New Year's Day of the previous year, the Zapatista National Liberation Army had emerged from the Lacandón Jungle, poorly armed and worse than outnumbered. The fighting was over in days, but the Zapatistas would prove agile, brilliant even, at using media and other symbolic means in lieu of actual weapons. Even the date of their rebellion, which coincided with the day the North American Free Trade Agreement took effect, was chosen for its symbolic impact. Their revolt was not just against the inequality, poverty, and repressive violence that had been suffered by indigenous communities since the arrival of Cortés, but against something called neoliberalism, of which NAFTA was only a small part.

Despite an expensive undergraduate education and voracious reading habits, that was a word I had never come across until I first encountered it in a Zapatista communiqué. It was not at that time a word we used in the United States. Even the cleverest fish rarely talk much about water. But there it was: the creed that guided my society had been named, its belief in the redemptive power of financial markets, in the distributive glory of unfettered capitalism, and in the irrelevance and even danger of any politics that sought to subject the latter to anything more than the most minor corrections. As its name implied, it was an eighteenth-century faith refurbished for the close of the twentieth, and its ubiquitous apostles spoke in a language—of progress and liberty and the triumphs of technology and instrumental reason—that their predecessors would have had little trouble understanding, a language that dominated the discourse so effectively that for a few decades they could safely insist that every other tongue spoken on the planet offered only gibberish.

I had arrived in the city of San Cristóbal de las Casas a few months earlier to work on a documentary about the Zapatista uprising. The filmmakers' finances kept falling through and the production was delayed, but I liked it there, and stuck around, sending the director occasional updates about the political situation while teaching English, studying Spanish, and befriending half the street kids who hung out in the plaza. They taught me a few words of Tzotzil, one of several local Maya languages, and laughed at my pronunciation. There wasn't always running water in the house my then girlfriend and I shared with a stoned and generally hostile English expat, but our share of the rent came out to fifty dollars a month. Every morning when I left the house I would look down the sloping streets at the city with its red tile roofs, the green mountains rising on all sides, the sky a depth of blue that I had never seen before.

That June, the Zapatistas had put out a call for a national referendum. They were struggling then, as they would be for years to come, to turn the short-lived uprising into a national movement. The referendum was one of many attempts to sidestep a political system that they saw as irredeemably corrupt, and to forge, in the process, an authentically democratic politics of resistance. It asked five yes-or-no questions. Among them: Should the Zapatistas convert themselves from a guerrilla army into an independent political movement? Should they join other already-existing organizations? Did everyone agree on their principal demands? My girlfriend and I joined thousands of other volunteers who helped administer the poll around the country. I remember long hours in fluorescent-lit church basements, a general air of urgency and excitement from which fear was not entirely absent. Though the fighting had ended, much of the state was still under the control of the military, and the Zapatistas and their supporters were being quietly harassed, intimidated, and sometimes murdered.

Except for the shantytowns at its periphery, where the kids in the plaza lived, San Cristóbal was a Ladino town, meaning that it was white and mestizo. The racial hierarchy was strict, and energetically enforced. It was analogous to the systems in place in the Jim Crow South. During the rainy season, the streets would turn briefly into rivers, so the sidewalks were built high and narrow. Until recently, I was told more than once, Indians had been expected to yield the sidewalks to Ladinos, to descend into the mud so their superiors could walk past them without dirtying their heels.

That day in August, the sidewalks were packed. The streets too. I remember climbing onto a roof so that we could see them streaming into town by the thousands, pouring in from every village around. Some came from places far enough out that they must have walked all night to personally deliver their ballots. Many were barefoot. They were short people but even the older ones who were bent from a lifetime of work in the fields looked very tall that day.

That's it. That's the story. They had a message to deliver: that they were not gone or dead or irrelevant, but alive, unbowed, and unafraid. For a few hours they gathered in the plaza and in the square outside the cathedral. They handed over their ballots, and then walked home.

≋

A foot of snow just fell on the East Coast, the fourth nor'easter in three weeks. The equinox was Tuesday, so it's spring now, officially. You have to dig deep to find anything in the news linking this to climate change, but the science is easy to track down. Increasing evidence ties extreme winter weather in the northeastern United States to rising temperatures in the Arctic. The low-pressure zone that locks cold air over the pole is collapsing as the Arctic warms. Also, the last male northern white rhinoceros on the planet died on Monday, in Kenya. The photos were unbearable: one of his keepers crouching to embrace him as he lay on the dirt-and-straw floor of his pen, his skin hanging like a rough blanket thrown over his giant frame.

Our Rhino survives. He's starting a trade war with China, and yesterday he went and did it, elevating a rabid, mustachioed cretin despised by more or less everyone who has ever met him to the post of national security advisor. Tune in tomorrow for another episode.

And the Great Pacific Garbage Patch is as much as sixteen times larger than previously estimated: 617,763 square miles of swirling plastic crap, larger than the combined areas of the states of California, Nevada, Arizona, New Mexico, and Utah.

Everything changes, even time. Hegel, lecturing in 1830 at the University of Berlin, described nature as dull and essentially static, "a tedious chronicle in which the same cycle recurs again and again." Yawn. It was in the more stimulating "theatre of world history," he promised, that "the spirit attains its most concrete reality." Time only started moving in a meaningful way, in other words, when humans were involved, and then only certain humans, on certain parts of the globe.

Africans, predictably, did not make the cut. Sub-Saharan Africa was, wrote Hegel in a passage that descends into breathtaking racism, "the land of childhood, removed from the light of self-conscious history and wrapped in the dark mantle of night." That is almost certainly the kindest thing he had to say on the subject. (He continued: "The negro is an example of animal man in all his savagery and lawlessness . . . to comprehend him correctly, we must abstract from all reverence and morality, and from everything which we call feeling.") Native Americans did not fare better. In Hegel's telling they were weak and passive, "obviously unintelligent individuals with little capacity for education." They could almost be said not to exist at all, and in any case possessed "a purely natural culture which had to perish as soon as spirit approached it." He meant that very literally—nearly all of the hemisphere's original inhabitants, he declared, had been "wiped out" not long after their first contact with Europeans. In 1830 this may still have been wishful thinking, but for Hegel, it was as it should be. History was not for such as they:

"Culturally inferior nations . . . are gradually eroded through contact with more advanced nations." Or not so gradually. Which left Asia, where "the light of the spirit first emerged," before it leaped to Europe, its proper home. Hegel is worth quoting again on that count: "World history travels from east to west; for Europe is the absolute end of history, just as Asia is the beginning." History had a direction in space as well as time. It belonged to Europe.

Until recently, most Europeans bore this burden lightly. Wars, famines, and plagues came and went, announcing little but more war or some fresh pestilence. Christ's return hung like a lantern at the end of time, and though some might pray that it hurry or be certain it was nigh, time nonetheless lazed past according to the rhythms of the plough and the scythe, the wheel of the saints' days and the opportunities for rest and carousal that accompanied them. For all but a minuscule literate elite, the years passed without number. Churches did not trouble to record the years of births, baptisms, and marriages until the sixteenth century. Common people's lives were ruled by neither days nor hours. The church bells would ring for matins and vespers, but no further demarcation of the time was necessary or even possible. Mechanical clocks only began to appear on church towers in the fourteenth century, and then only in cities and large towns. For the next four hundred years, it would have made little sense to slice an hour into smaller bits. Most clocks did not even have minute hands. If the need arose, time could be subdivided into commonly understood and distinctly human-sized quantities—the time it takes to mow a field of grain, say, or to recite an Our Father, or to urinate. An interval known as a "pissing while" made it into Shakespeare's *Two Gentlemen of Verona*, as in: "He had not been there—bless the mark!—a pissing while, but all the chamber smelt him."

Over the course of the seventeenth century, more accurate

clocks began to proliferate, and eventually to migrate from town squares and cathedral belfries into the drawing rooms of the gentry. In another fifty years, technology—specifically, the development of the balance spring and regulator—allowed clocks to miniaturize without any sacrifice in accuracy, and to take refuge in the pockets of an emerging class of merchants and manufacturers. In 1687, ten years after the invention of the balance spring, Isaac Newton elevated time in an extraordinary fashion, defining it as an absolute and independent force that progressed at a consistent rate regardless of who was perceiving it, or if it were perceived at all. Even if the universe were entirely empty, time would tick on, echoing through the void like God's very pulse. And why not? Any mortals with sufficient income to spare could by then carry time around with them, trapping it in a compact silver case, embossed perhaps with golden stars. Whether its owner was awake or asleep, alive or dead, it would tick and tock for eternity. Unless they neglected to wind it. By 1760 the personal timepiece had become sufficiently widespread that a British satirist could complain that "Sir, will you have your clock wound up?" had become "the common expression of street-walkers."

Over the course of the next century and a half, time, once the property of none and all, would belong to the rich and the powerful—first to the factory owners, then the railway magnates, then to the politicians and military men—divvied up according to the needs of capital and conquest. Technology, the British historian E. P. Thompson observed, was never the decisive factor. Though the key horological innovations had occurred decades earlier, the "general diffusion" of clocks and watches in England did not take place until the 1790s, "at the exact moment when the industrial revolution demanded an exact synchronization of labour." Large-scale industrialization, Thompson argued, had reconfigured labor. Workers who had once bartered their skills or their strength now hired out their *time.* They were paid by the day or the hour rather than the task or piece.

In a very literal sense, time became money. It became something wholly abstract, something independent of our bodies, of the circulation of our blood and the transit of our thoughts and desires. It could be used or misused, spent or wasted, even stolen. Moments of inactivity were suddenly transformed into something that would have been nonsensical a few years earlier: a "theft of time" that vigilant managers strived to guard against. The ironworks in Swalwell, in northeast England, which was for a little while the largest factory in Europe, was the first to employ a monitor tasked with noting every worker's arrivals and departures to the minute, and with guarding the clock to prevent workers from setting it forward or back. Power over time had become power over labor. Which is to say, over people.

It was at the same moment—which was also the moment of revolutionary upheaval in France and in Haiti, of Condorcet's frenzied ode to human progress, and, by many measures, the beginning of the modern era—that with the development of the coal-powered steam engine, the industrializing societies of Europe and North America began releasing sufficient carbon dioxide to eventually alter the climate of the earth. It was then too that the stars began to disappear from the skies of the cities and towns of Europe and North America, and from the consciousness of their inhabitants, smeared away by the smoke of all those factories and by the glow of the gas lamps that in the early decades of the 1800s were beginning to illuminate the streets of European capitals.

It was then that time got inside us. Schools for the poor began to teach "time thrift" and to discipline children into "the habits of industry." Workers, wary of having their pay docked for tardiness, formed a steady market for newly inexpensive, mass-produced watches. Time ticked away in their vest pockets, closer to their hearts than their children or their lovers. (After the Battle of Little Bighorn, the Oglala Sioux visionary Black Elk recalled pulling "something bright" from the belt of a fallen U.S. cavalryman: "It was round and bright and yellow and very beautiful and I put it on me for a necklace. At first it ticked inside, and then it did not anymore. I wore it around my neck for a long time before I found out what it was.") For workers too, time—the very substance of a life—had become money. Leisure, and the infinite forms of joy it can engender, was refigured as waste. Head-on resistance to this new "time-sense" was rare. "Workers began to fight," Thompson wrote, "not against time, but about it." The great labor battles of the nineteenth and early twentieth century were waged over the length of the working day— limiting it to twelve hours, then ten, and finally eight. The measure of a worker was no longer his expertise or his vigor, but how fully he had surrendered to the disciplines imposed upon his *use* of time, and "perhaps also by a repression, not of enjoyments, but of the capacity to relax in the old, uninhibited ways."

Time's dominion was nonetheless incomplete. The hour might be uniform within a single factory or mill, but every municipality in Europe and the United States still kept its own time, setting the town clock to twelve when the sun hung at its highest. Newton's *Principia* notwithstanding, time was anything but absolute. In 1857, when it was noon in Washington, D.C., it was 12:08 in New York City and 11:48 in Pittsburgh. Until the railroads began to connect these far-flung cities, no one had any cause to care. But trains had to keep a schedule, and to avoid each other on the tracks, so the railroads took charge of regulating the flow of time. In England, the Great Western Railway solved the problem by announcing, in 1841, that its trains would run on London time, also known as Greenwich Mean Time, wherever between London and Bristol they might happen to be. (In *Dombey and Son*, Dickens bemoaned the expanding powers of the "monster train": "There was even railway time observed in clocks, as if the sun itself had given in.") Frequent travelers purchased watches with two minute hands—one for the local time, another for the railroad's. In 1847, the secretary of the Liverpool and Manchester Railway appealed to nationalist sentiments while calling for a radical standardization of time: "There is sublimity in the idea of a whole nation stirred by one impulse; in every arrangement, one common signal regulating the movements of a mighty people!"

Across the Atlantic, the complexities were compounded by the enormity of the continent. By 1882, competing American railways ran on fifty-three different times, not counting the local times of the stations that they served. The major railroads organized a General Time Convention, which, with encouragement from the U.S. Signal Service, then part of the Department of War—the military had its own reasons to favor synchronization—managed on November 18, 1883, to establish a uniform, coast-to-coast Standard Railway Time, broken into the four time zones still in use today. (It was earlier that year that the Southern Pacific Railroad had completed a new route connecting New Orleans to the West Coast, for the maintenance of which the railroad had deprived the Chemehuevi and Serrano of access to the springs at the Oasis of Mara.) Noon fell twice in New York that day, once at the usual hour and then again four minutes later, according to the newly christened Eastern Standard Time. "The citizens did their share of the work by carrying the new time to homes, hotels, and stores," *The New York Sun* reported the next day. "Everybody who had it, or who even thought he had it, distributed it among his friends, so that by 6 o'clock in the evening, most wide-awake people had it in their houses or their pockets." Within a year, most of the country had followed the railroad's lead.

Other innovations would tie the world in a smaller knot still.

In 1876, Sandford Fleming, chief engineer for the Canadian Pacific Railway, published a pamphlet titled *Terrestrial Time*, in which he argued for the creation of a single global time standard. The telegraph and the steam locomotive, he wrote, had already "rendered the ordinary practice of reckoning time but ill suited to the circumstances which now exist." And they had hardly begun—"We may rather assume," Fleming went on, "that they will still achieve greater triumphs in the work of colonization and civilization." Australia and Africa, he predicted, would soon "be pierced, perhaps girdled by railways," and "Asia, with more than half the population of the world, must in due time yield to the civilizing pressure of steam and participate in the general progress." The solution he proposed was to divide the earth into uniform time zones, one for each hour of the solar day. To illustrate this notion, he imagined the globe bisected at the equator by a plane and split into twenty-four bands of equal width. At the center of this plane, like the dial plate of a watch, he imagined a minute hand and an hour hand. The latter would "rotate from east to west, with precisely the same speed as the earth on its axis." The planet, so envisioned, had become an enormous clock, its workings concealed in its core.

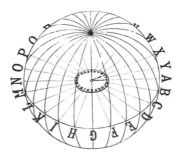

Fleming would spend the next few years proselytizing for the establishment of "one standard time common to all peoples throughout the world," whether they wanted it—or even knew about it—or not. The rub was where, exactly, to place the prime meridian. From a scientific perspective it makes no difference which vertical hoop around the planet marks the zero hour, but the British, who already controlled most of the world's trade, favored one passing through Greenwich, the French though Paris, the Americans Washington. In October 1884, representatives of twenty-six countries—"most of the nations of the earth," per the U.S. secretary of state at the time—gathered in Washington to select a meridian that would be used for reckoning longitude as well as a single, planetary time. The delegates were naval officers, railroad men, engineers, astronomers, and diplomats. Fleming, who was there representing the British colony of Canada, marveled at the mystical qualities of the object under debate: "When we examine into time in the abstract, the conviction is forced upon us that it bears no resemblance to any sort of matter which comes before our senses; it is immaterial, without form, without substance, without spiritual essence. It is neither solid, liquid, nor gaseous."

He recovered quickly, taking refuge in the prevailing Newtonian dogma: "Of one thing there can be no doubt. There is only one, and there can only be one flow of time, although our inherited usages have given us a chaotic number of arbitrary reckonings of this one conception . . . the progress of civilization requires a simple and more rational system than we now have." Over the course of a month, the delegates fought out the details. The French consul general, despairing of imposing Parisian time on the world, demanded the selection of a "neutral" meridian that "should cut no great continent—neither Europe nor America." (No one was there to argue with him: the remaining continents were represented only by Russia, Turkey, Hawaii, and Japan.) But the globe had already been so extensively divvied out among colonial empires that no neutral ground could be found. Even the Bering Strait, an American representative speculated, might not remain unclaimed for long: "Who knows when America will step over and purchase half of Siberia?"

Eventually, the delegates—with France abstaining—bowed to imperial realities. The prime meridian, and the basis for all terrestrial navigation, would pass directly through the Royal Observatory at Greenwich, on the south shore of the Thames, within the city boundaries of London. Time would begin, and end, in the capital of the most powerful empire on earth. Each new day would henceforth commence at midnight, Greenwich Mean Time. The French delegate, who surely would not have objected had Paris been chosen, appeared to relish his role as killjoy. "Science appears here," he complained, "only as the humble vassal of the powers of the day to consecrate and crown their success.

"But, gentlemen," he went on to warn, "nothing is so transitory and fugitive as power and riches. All the great empires of the world, all financial, industrial, and commercial prosperities of the world, have given us a proof of it, each in turn." Time has, as they say, proven him correct. The vast colonial empire controlled by the British would collapse in fewer than eighty years. Time, if little else, would still be theirs.

It had been months, I realized, since I'd seen any friends, so I drove to L.A. for the weekend. I went first to Echo Park for a coffee with B. He and his wife live across the street from the apartment where I lived for eleven years. It was haunted—literally: there had been a murder-suicide in my bedroom a few years before I moved in, and a ghost occasionally showed itself to a houseguest. The elderly owner and her still-older brother, who lived in the apartment next door, were both hoarders. Every year or two the city inspectors made them clear a path through the crap on the porch, but apparently they died because the house had been sold recently and the porch was filled with potted plants instead of broken TVs and toilet bowls and cast-away toys. Whatever the rent is now, I imagine the ghosts have not moved out.

I had almost forgotten about the bougainvillea, how bright they are, explosions of pink climbing the stucco walls, and how soft the light is in the afternoon, how it makes the colors seem magically deeper and it really does feel like anything is possible. For a long time, that light was enough to sustain me there. I took the Harbor Freeway down to Long Beach to see some friends. I spent the night on their couch, and in the morning I woke before anyone else and drove down to the beach for a run. I wanted to see the ocean, to smell it, and to listen to the surf roll in. When I got there I couldn't smell a thing. The breakwater around the harbor had tamed the waves to nothing and the oil platforms and cargo ships blocked most of the horizon, but there it was nonetheless in all its patient blueness. Later, when I had showered and was sitting at the kitchen counter drinking coffee, my friends' twelve-year-old son C. emerged from his room in a blue bathrobe printed with ringed planets and stars. He asked me when I was going to write a new book. I told him I had started one already. He asked what it was going to be about and I did my best to explain. He told me there was a Greek god named Chronos that I should look into who could make time slow down or speed up. He could even stop time, if he wanted.

After breakfast I drove to Alhambra to meet S. and D. We hiked up to the top of Echo Mountain in the national forest north of the city. The trail was crowded but rain earlier in the week had cleared the air and we could see the sun shining off the ocean way out at the end of the long, straight boulevards, the glass towers of downtown Los Angeles rising like a castle keep from the low mess of gridded streets. At the top of the trail were the ruins of the White City. At the turn of the century it had been a grand resort with tennis courts and a dance hall and its own little zoo. All of it had been connected to Pasadena by rail, the tracks ascending the steep face of the mountain. There were just a few rocks there now, the outlines of the foundations.

A little farther up, we found the ruins of the observatory. Thaddeus Lowe, the autodidact former chief aeronaut of the Union Army Balloon Corps, inventor of the ice machine and of his own patented process for the production of hydrogen gas, had built the resort and the railway and in 1893 began paying an astronomer to live on-site and deliver lectures to the public about the wonders of the heavens. By 1905 every building but the observatory had been destroyed by fire, wind, and floods. In 1928 a storm tore away the dome, blowing away the entire building with the resident astronomer, the third to hold that post, crouched inside to save the precious lenses. All that's left now is a hexagonal foundation and the chimney-like stone tower that once supported the weight of the telescope. "I don't know why they thought anything would last up here," laughed S., his foot resting on the stones.

We talked about bears and S. and D.'s daughter in Alaska, about the smallness of literary politics and the anti-gentrification battles in Boyle Heights, how they had fallen into sectarian skirmishes like back in the seventies all over again. D. told me she visits her mom a few times a week and after dinner they sit and drink beer and watch the news in Spanish on Univision. Every night it's the same thing, she said, more ICE raids and people saying don't call the police for anything or even the fire department because ICE might come with them and take away your dad or your mom or your kids. When we got back to the house she made taquitos and a salad, and sautéed corn, squash, and nopales with garlic and the smoky paprika I brought them from Spain last year. S. talked about his students up at CalArts and how incurious they are, how disappointed he had been by their reactions, or lack of reactions, to the Kenneth Rexroth poem he had assigned the week before.

They had both left for work when I got up this morning. I poured myself what remained of the coffee and found *The Complete Poems of Kenneth Rexroth* propped up against the chimney by the door. I looked up the poem S. had taught that week, "The Signature of All Things." It's named for a work by the seventeenth-century German mystic Jacob Boehme. Rexroth reads it in the first stanza while lying beside a waterfall, watching the leaves and the sunshine tumble down around him: "The saint saw the world as streaming / In the electrolysis of love." Light drifts in fragments throughout, scattered bits of divinity glimmering in through everything. At the end, the poet cuts up a rotten log and leaves it in the sun to dry for kindling. That night he steps out on his porch and looks up, "At the swaying islands of stars." He sees something gleaming beneath him too, phosphorescence glowing in the wood he had chopped, everywhere, above and below, "scattered chips / Of pale cold light that was alive."

On the drive back to Las Vegas I passed the solar-thermal plant again. On sunny days, I had read, its mirrors heat the air to temperatures as high as one thousand degrees Fahrenheit, killing as many as six thousand birds a year. Workers at the plant call them "streamers": they ignite in flight and streak smoking to the ground. The sun was high this time and the towers were blinding bright with all the fragmented and reflected sunlight bouncing onto them and concentrating there. Around them on the desert floor the mirrors trapped little chips of sky, hundreds of them, every one of them blue and filled with clouds.

≈

It was a man named Samuel Pierpont Langley, chief astronomer of Pittsburgh's Allegheny Observatory, who innovated the *sale of time*—not any individual worker's time, but time itself, delivered by telegram twice daily. The observatory, which did not have a mountaintop resort to bring in revenue, had fallen into debt, so Langley established a subscription service whereby local jewelers could pay to receive the time, astronomically determined, and accurately set the clocks and watches in their showrooms. In 1869 the Pennsylvania Railroad, soon to be the most-traveled railroad in the United States, became the first of forty-two railways to sign on. From its conception, standardized time was privatized time. When the heads of the major U.S. railways later agreed to divide the country into uniform time zones, it was a signal from the Allegheny Observatory that marked the first noon of Eastern Standard Time.

With the income he brought in selling time subscriptions, Langley was free to fund his more ambitious research on the effects of solar radiation on the earth's atmosphere. In 1880, he invented the bolometer, a device capable of measuring electromagnetic radiation. With it, after years of observation, he was able to estimate the temperature on the moon and to determine that all prior estimates of the absorptive capacity of the earth's atmosphere had been far too low. The data Langley compiled would allow the Swedish chemist Svante Arrhenius to more precisely calculate the degree to which the sun's energy is absorbed by the carbon dioxide in the atmosphere. If coal-powered industrialization continued to the point that carbon dioxide emissions doubled, Arrhenius predicted in 1896, the surface temperature of the planet would increase by between five and six degrees centigrade. (By that time Langley had taken over the helm of the Smithsonian, which means that the *Fourteenth Annual Report of the Bureau of Ethnology to the Secretary of the Smithsonian Institution,* in which James Mooney's findings on the Ghost Dance were published, was addressed directly to him.)

Arrhenius, whose perspective may have been warped as much by a lifetime of Scandinavian winters as by unbending faith in progress, regarded the prospect of planetary warming as yet another perk of human ingenuity. Should the earth's atmosphere heat up a few degrees, he remarked, "We would then have some right to indulge in the pleasant belief that our descendants, albeit after many generations, might live under a milder sky and in less barren surroundings than is our lot at present."

Maybe I'm doing it wrong, reading all these books and writing every day. This, from Barbara Tedlock's *Time and the Highland Maya*: "Understanding is achieved by a combination of several actions: the mixing, grabbing, and arranging of piles of *tz'ite* seeds; the counting and interpreting of the 260-day divinatory calendar; and the jumping and speaking of the diviner's blood."

Time's newfound sovereignty was not universally welcomed. The workers who stooped beneath the strictures of the new regime rarely had the opportunity—or indeed the time—to record their protests for posterity. The writers, though—because a writer is above all a person who *makes the time* to write—complained bitterly of its oppressiveness. Some of them anyway. Charles Baudelaire, the great and bilious poet of night and crowds, the gutter and the boulevard, included a poem called "The Clock" in the second edition of *The Flowers of Evil*, published in 1861. "Clock!" it begins, "Sinister god, appalling, unperturbed . . ." In six brief stanzas, Baudelaire portrayed time as an evil deity, an archer shooting sorrow into human hearts, an insect sucking up life through its "filthy proboscis," a sort of metal-throated robot, and a gambler who never loses and never has to cheat. The clock, which elsewhere was standing in for progress and the inevitability, strength, and power of the railroad and the steam engine, became for Baudelaire a memento mori, a throwback to the reaper's hourglass.

Edgar Allen Poe's stories and poems also associate clocks with murder, torture, death. (Baudelaire, who introduced Poe's work to the French, was conscious of their kinship.) Think of "The Tell-Tale Heart," where the too-clever killer is given away by the still-beating heart of his victim, which thuds on with "much such a sound as a watch makes when enveloped in cotton." Or "The Pit and the Pendulum," in which the narrator, imprisoned by the Inquisition, awakens in his dungeon to see the "figure of Time as he is commonly represented" painted on the ceiling above him, except that instead of a scythe Time holds "a huge pendulum, such as we see on antique clocks." This one has a sharpened edge and is descending, one swing at a time, closer and closer to the narrator's heart. Time and Death are almost interchangeable.

I could go on for pages about Poe: "The Colloquy of Monos and Una" with its dark judgment of the "infantine imbecility" of the notion of progress, the comic "A Predicament," in which Signora Psyche Zenobia gets stuck in a clocktower and is decapitated by the sharpened minute hand, but continues to narrate the story, even delivering a speech, which her body "could hear but indistinctly without ears." But Baudelaire, whom Benjamin adored, also wrote, in *The Flowers of Evil*, a poem called "Owls." It's a sonnet: four short lines, then four short lines, then three and three again. His owls do not fly. He does not say how many there are, but for the length of the poem, they never leave the tree. They perch there, neatly arrayed, "like strange gods." Or "foreign gods," depending on which translation of *étrangers* you prefer. (In her version, Edna St. Vincent Millay went with "eastern gods," believing, I suppose, that the Orient offered a successful fusion of both connotations.) If Hegel found the owl's wisdom in nocturnal flight, Baudelaire found it in stillness: "Without stirring they will remain, up to the melancholy hour / when shadow, pushing the sun aslant, takes over." The restless human ambition that prods history forward is not a virtue here. It's a curse. We are all "drunk on passing shadows," chasing the illusion that things are better elsewhere, or will be, or could be, stooping always under the punishment we bear for "having wanted a change of place."

Owls are smarter. They don't fly until they have to, until it's night.

*

Last night I dreamed of Walter Benjamin. I don't remember much about the dream, or anything, really, other than a certain mist of uneasiness. It still hasn't lifted. I have a bad cold and haven't gone out much or written anything for days. In Washington the usual theatrics. On Saturday a headline in *The Times* ended with the sentence "No One Knows What Comes Next." Some quorum of editors considered this worth printing. How uncertain must uncertainty be before it is counted as news?

*

I can think of at least one exception to E. P. Thompson's generalization that workers in the nineteenth century did not fight "against time, but about it": the much-maligned anarchist tailor Martial Bourdin, who died shortly after attempting to blow up the

Greenwich Observatory.* Other than the sensational circumstances of his death, next to nothing is known about him. He was by most accounts a taciturn man, and those who knew him well had every incentive to stay quiet, or to lie. We know that Bourdin was born in France, in Tours, in 1858, and at the age of sixteen was jailed for two months for the crime of "having attempted . . . to organize a meeting." Three or four years later he emigrated to London at the invitation of an older brother, Henri, who had been active in London anarchist circles at least as far back as the 1880s. The brothers worked together as dressmakers, along with Henri's wife and her sister, out of a workshop in central London, not far from where the BT Tower now stands, a short walk from the room Bourdin rented on Fitzroy Street.

By February 15, 1894, business was apparently slack. Bourdin was out of work. That afternoon, "respectably dressed," per the next day's *Times*, he walked to Westminster and caught a tram to Greenwich. He got off at the last stop, asked the conductor for directions to Greenwich Park, and was seen entering the park shortly thereafter, carrying a parcel. At 4:45 p.m., two employees of the Greenwich Observatory, a Mr. Thackeray and Mr. Hollis, heard what they later described as a "sharp and clear detonation, followed by a noise like a shell going through the air." They rushed outside and saw a man crouched on the path just below the observatory. The bomb Bourdin had been carrying had blown off his hand, torn a wide and jagged hole in his abdomen, and scattered bits of gore to a distance of nearly sixty yards in the direction of the observatory wall. He was

---

* Actually, I can think of one other: in his "Theses on the Philosophy of History," Benjamin mentions that when workers in Paris rose up against the monarchy of Charles X in July 1830, they "simultaneously and independently" began shooting at public clocks around the city.

sufficiently lucid to pronounce the words "Take me home." They took him instead to the Seaman's Hospital, a half mile away, where he died within the hour.

Bourdin's intentions were not at all clear, but the "narrow, zigzag, and secluded path" on which he was found, *The Times* pointed out, "leads practically nowhere except to the Observatory." Several letters, a small and broken bottle of sulfuric acid, and thirteen British pounds in gold, a considerable sum at the time, were found in Bourdin's pockets, but no further clues to his motivations. As is often the case, the truth would be useful to no one. Britain at the time had the most liberal asylum laws in Europe. Foreign-born radicals and dissidents were able to live in London under relatively little scrutiny from the state. Not everyone was pleased with this. "The 'right of asylum,' may be a sacred thing, but we can have too much of it," huffed the conservative *St James's Gazette* two days after the bombing. Perhaps this sounds familiar.

Other newspapers fanned the flames, rushing to speculate that Bourdin's attempt on the observatory had been but one attack among many planned by an international anarchist cabal, "the most desperate and dangerous of any revolutionary plot that has ever had its headquarters in London." Police sources initially told the press that there was "practically no room for doubt" that the explosion was an accident and that Bourdin, knowing that his arrest was imminent, had been attempting "to rid himself safely of the explosives" when he stumbled over a root. Politics appears to have intervened. At a coroner's inquest several days later, the queen's chief inspector of explosives testified that he could "arrive at no other conclusion" but that Bourdin's intention was to act "against the Observatory, or its contents, or its inmates." This would become the official version.

For their part, British anarchist circles, eager to preserve the relative ease of their existence in England, distanced themselves from

Bourdin. One prominent anarchist editor published a pamphlet tarring him as the dupe of an agent provocateur in the pay of English authorities. Nine years after Bourdin's death, Olivia and Helen Rossetti, who as teenagers had founded the radical journal *The Torch*, expanded on this version in the pseudonymous novel *A Girl Among the Anarchists*, in which "an obscure little French Anarchist" blows himself up in a London park. The observatory doesn't rate a mention. The Rossettis' cousin, the novelist Ford Madox Ford— they were also the nieces of Dante Gabriel Rossetti, of *Lady Lilith* and "Eden Bower" fame—later alluded to the bombing in passing to his friend Joseph Conrad, dismissing Bourdin as "half an idiot."

Be careful what you say to a writer: Conrad would make the anecdote the basis of a novel, *The Secret Agent*, published in 1907. In his version the anarchists—and pretty much everyone else—are foul, cynical creatures, motivated more by sloth, vanity, and greed than by anything like a principle. The unfortunate Bourdin would be transmuted into Stevie, "poor Stevie," dumb and innocent, with a squint and a drooping lip, fooled into carrying out an attack on the observatory by his grotesque and unscrupulous brother-in-law Verloc, an agent in the pay of an unnamed but reactionary Eastern European state. "What is the fetish of the hour that all the bourgeoisie recognise—eh, Mr Verloc?" asks Verloc's employer, the sinister Mr. Vladimir, and immediately answers his own question. "'The sacrosanct fetish of to-day is science . . . Yes,' he continued, with a contemptuous smile, 'the blowing up of the first meridian is bound to raise a howl of execration.'"

In an "Author's Note," published thirteen years later, Conrad, the great psychologist, played coy, describing Bourdin's final act as incomprehensible, "a bloodstained inanity of so fatuous a kind that it was impossible to fathom its origin by any reasonable or even unreasonable process of thought." It is a curious admission. In his other

novels, Conrad was able to parse even the subtlest and most para-
doxical aspects of human behavior. But in *The Secret Agent*, it was
important to him to maintain that Bourdin's "outrage could not be
laid hold of mentally in any sort of way; so that one remained faced
by the fact of a man blown to bits for nothing even most remotely
resembling an idea." Stevie had to be a fool, Verloc a scoundrel,
Mr. Vladimir a devil in blue silk socks. What could not be admitted, by any means, was the possibility that a sane man, ruthless in
his idealism, might want to wage war against Time.

I neglected to make any plans for the weekend and ended up staying
in the apartment for almost twenty-four hours before I realized I
needed to get out. I'm not counting a stroll to the mailbox and a
coffee on the stoop. I drove to Henderson, to a running trail paved
alongside a wash that snaked through the suburban cul-de-sacs. I
could see them behind the fences, identical houses, identical drive-
ways, identical streets, but the desert willows along the wash and the
blooming jasmine in the backyards perfumed the air, and I didn't
mind the view. I checked Twitter before I went to bed that night
and found myself scrolling past image after image of what looked
like sleeping children, their faces a strange gray-blue. A chemical gas
attack on Dhouma, east of Damascus, one of the few areas still held
by Syrian rebels. I didn't click the links, didn't want to see it, didn't
want to know.

Later I went out for Chinese noodles with T., the other fellow
at the Institute here. She's just back from her book tour, so we cel-
ebrated with dessert at one of the yuppie places downtown: bread
pudding made with donuts from the artisanal hipster donut shop
next door. I dropped her off and spent the next hour sitting up
in bed staring at my phone. There had been a missile attack on a

Syrian airbase not far from Homs. The same base that Israel struck in February. Fourteen killed this time, several of them Iranians. A tweet from the AP did not reassure: "BREAKING: U.S. officials: The United States is not carrying out airstrikes in Syria." Then more reports: Israel was bombing Gaza, for no reason that made sense. Now Syria and Russia are saying it was Israeli F-14s that struck the Syrian base. Israel has not denied it. This morning the Rhino announced he would decide how to respond soon, probably by the end of the day. If Russia was responsible for the Dhouma attack, he said, "it's going to be very tough, very tough. Everybody's going to pay a price."

≈

Bourdin's choice of target put him in curious company. In 1910 Sir Robert Anderson, a lifelong veteran of the British secret political police, published a memoir with the jaunty title *The Lighter Side of My Official Life*. On the day of the explosion in Greenwich Park, he revealed, "information reached me that a French tailor named Bourdin had left his shop in Soho with a bomb in his pocket." The shop was in Fitzrovia, but regardless, Sir Robert complained, all he could do was dispatch officers to sites he thought an anarchist might be likely to attack, and wait. Bourdin's actual destination, he wrote, "was the last place the police would have thought of watching."

"In war the guns of an enemy would no doubt spare an astronomical Observatory, for none but *savages* would wish to injure an institution of that kind."

It's a nice thought anyway. Between 1943 and 1945, British and American bombers destroyed observatories in Munich, Dresden, Leipzig, and Königsberg, but Sir Robert was by then long dead.

≈

There's an owl in the middle stanza of that Kenneth Rexroth poem, the one about Jacob Boehme and the chips of living light. It's a little one and it flies, Rexroth wrote, "on wings more still than my breath." It lands above him, on a branch. He's not reading anymore at this point. He's standing at the edge of a forest, alone. The moon is full. The poet shines a flashlight up the tree trunk and catches the owl's eyes in the beam. It stares back at him and cocks its head, as curious as he is.

I keep planning to drive outside the city just to see the stars. I'm afraid that I'm forgetting them, not only their names and relative locations but the sense of perspective that they force on you. I've lost that already, or most of it. I wouldn't have to go far. Thirty of forty miles should be enough to escape the lights. An hour in the car round trip, less time than I spend driving to Trader Joe's and back at rush hour. Instead I stand in the yard, sticking to the concrete because in the dark I can't tell where the dog turds are on the Astroturf—the neighbors' dogs roam free, and every day brings fresh deposits—grumbling up at the few lone blurry lights up there, cursing the Luxor and its stupid, giant light beam, complaining on the phone to L. that I don't know if I'm looking at Spica or Arcturus, Sirius or Capella. And then I go back inside, and shut the door, and read, and write some more.

❧

For the astronomers of ancient Mesopotamia, the stars that crossed the sky each night were not just distant lights of fixed path and varying brilliance. They were letters, words, writing, a script that could be read. Babylonian inscriptions from the first millennium

B.C. referred to the star-filled sky as "the heavenly writing." (And how bright its verses must have been, unmarred by smog or the glow of a single neon bulb, no Luxors anywhere on earth.) In the Sumerian language, the same word that meant "star," *mul*, also referred to cuneiform signs inscribed on tablets of clay. Stars were writing and writing was stars. Astronomical observation—the tracking of the paths of the stars and planets and the prediction of celestial events—was not an abstract science but a form of reading, an attempt to transcribe and interpret the scribblings of the gods.

This conception, of course, could be no older than writing itself. Which is to say no more than about five thousand years old. So the consensus has it. The most ancient archaeological evidence of what most scholars recognize as writing—petroglyphs don't count—is a cache of inscribed clay tablets dated to about 3100 B.C. and unearthed in the ruins of a temple to Inanna, the goddess of love and war later known as Ishtar. They were found in Uruk, possibly the planet's first city, in what is now southern Iraq. (The name *Uruk* is the ancestor of the modern *Iraq* and derives from a Sumerian word meaning simply "city"—it was likely the largest human settlement in the world at the time, with a population in the tens of thousands.) Most of the Uruk tablets appear to have been something like receipts: lists of goods and expenses, tallies of slaves, sheep, wool, prisoners of war, textiles, and grain. One is a list of ranked professions, from king and high officials down to gardeners and cooks. If these tablets count as the earliest form of writing, then the roots of written literature are not only unromantic but repressive: writing then emerged as a means of record-keeping, the contrivance of a society that had grown sufficiently bureaucratic and hierarchical to require a storehouse of memory that would transcend any of its individual members, a tool of social and economic control for a state eager to extract labor and tribute from its subjects.

But it was these finds, and others like them—later tablets would record the names of kings, the deeds of gods and mythic heroes, astronomical, medicinal, and botanical knowledge, instructions for the interpretation of dreams and omens, a creation myth very similar to the one that would open the book of Genesis—that convinced Europeans of the last three centuries that the origins of civilization, *their* civilization, must lie in Mesopotamia. This is what sent adventurers like R. C. Thompson scrambling through the deserts of Iraq, snatching up every tablet they could find.

History, we are told, begins with writing, a convenient sleight of hand that eliminates most humans who have ever lived from inclusion in the grand parade. People without writing—or something Europeans could recognize as such—"remain involved in the obscurity of a voiceless past," said Hegel. "Not partaking of this element of substantial, veritable existence, those nations . . . never advanced to the possession of a *history*." Narrating the development of writing would become another way for Europeans to imagine the world as a vast mirror in which they could watch their own supremacy take shape. If human progress could be measured in the three-step ascent from savagery through barbarism to civilization, each stage had its analog in the development of writing: from pictogram to ideogram to the perfection of alphabetic script. As Jean-Jacques Rousseau put it in his *Essay on the Origin of Languages*, "The depiction of objects suits savage peoples; signs of words and propositions barbarous peoples; and the alphabet civilized peoples."

More than the rise of cities, codified laws, or even monotheism, phonetic writing, and particularly the alphabet, would become the measure of human progress. Rousseau may have had mixed feelings about the latter, but he would not have argued with the general frame or disagreed when Condorcet declared that it was the development of alphabetic script that marked the border between savagery and civilization. Only when words were spelled out in letters that represented nothing more than sounds did thought become sufficiently abstract that Europeans could recognize it as their own. Only once the linguistic sign was thus unchained from its object could representation and reality be distinguished as separate realms. Exploring the gulf between them would become the task of Western philosophical inquiry from Plato onward. It's a funny thing to boast about, that it was this gap, this wound, this fundamental alienation inserted into the very possibility of knowledge, that made the West self-consciously the West.

The cuneiform script that developed in Mesopotamia over the course of more than three thousand years—it remained in use into the early centuries of the first millennium A.D.—included aspects of all three of Rousseau's stages. Those earliest tablets from Uruk were pictographic. The sign meaning "head" looked like a human head in profile, with recognizable eye and nose and chin. The sign meaning "bread" looked like a wedge of pita, and the sign for the verb "to eat" combined the signs for "head" and "bread." Not all of them were quite so literal. Meaning extended outward. A sign based on an image of a foot came to represent not just the physical limb but the act of walking and standing as well as something you might stand on: the ground, the earth, even the generalized notion of a foundation or a place. "By these various shifts, each character could be enriched by an entire *constellation* of meanings," wrote the French Assyriologist Jean Bottéro. With time, the signs simplified until they were no longer recognizable as images. They became ideograms, characters rather than pictures, and increasingly included phonetic elements, signs that represented sounds as well as things, tying writing to spoken language as an intermediary between the sign and the object it strived to represent. By the end of the third millennium, cuneiform had become fully phonetic: each sign represented the sound of a spoken syllable, which could be combined with others to represent any word or name that could be pronounced aloud. It was syllabic rather than alphabetic and, by Rousseau's and Condorcet's schemata, more perfect, but not quite there.

This, though, is where things get interesting. The newly phonetic cuneiform signs retained their older pictographic and ideographic meanings. Each sign henceforth represented both a sound and a thing, and not just one thing, but the entire web of meanings that orbited the original image, the "semantic constellation" that it still carried within it like a chain of ancestors, or a crowd of ghosts. Or like the branches, leaves, and roots of a creosote, splitting off from the lost-but-still-present original. "Even after the invention of phonetism," Bottéro wrote, "the cuneiform writing system never abandoned its original, deep-seated habits of immediate reference to things . . . The name was inseparable from the thing." Written language was not distinct from but contiguous with material reality. The "written name, equal to the thing, constituted a material given, which was concrete, solid, and comparable to a substance of which each portion, even the smallest one, contained all the faculties of the total, just as the smallest grain of salt has all the characteristics of the heaviest block."

It was in this sense that Mesopotamians understood the universe as written: not just the stars, but everything in the observable cosmos was buzzing with meanings, whole constellations of them inscribed there by the gods. If the multiple and overlapping meanings of a name could be exhaustively unraveled, the past and the future—the destiny—of the thing so named would become clear. It was just a question of reading it well. Professional diviners deciphered the messages hidden in the entrails of sacrificed animals and in the shapes that oil made when poured over water, but also in the patterns and anomalies of the weather, the appearance and behavior of animals and men, the contours of the earth, the plots of dreams, of course the stars. As the archaeologist and art historian Zahrab Bahrani put it, "Their whole world was seen as ominous." Not ominous as in inauspicious or threatening, but ominous as in thick with omens, with meanings woven into the fabric of things.

The entirety of existence was a text waiting to be read. Which means there could be no line between the reader and the written. You, who are reading this, you too are written, you too can be read. And I, a writer, am already written through and through. Everything between us, everything that separates us, mountains, stars, years, shimmering thoughts and dreams that die with waking, all of it is a single chain of signs that do not point to another reality, only to this one, all at once.

There's a new mattress outside. It's smaller, just a twin-sized foam pad, the egg-carton kind, with those little peaks and spikes. It's right where the old one was, pushed up against the fence opposite my patch of yard, beneath the bamboo. I noticed it this evening as I pulled up from the alley because there was a man sleeping on it, or getting ready to sleep. It was early. The sun had just set. It's after ten now and he's still there, less than forty feet from the bed in which I'm sitting, propped up on pillows, typing this.

⁊

Obliquely, at least, Claude Lévi-Strauss wrote about Sumerian tablets too. In a much-quoted aside to *Tristes Tropiques*, the famed anthropologist halted the meandering flow of the book's narrative to take up, and just as quickly drop, a surprising argument about the activity in which he was engaged. Which is to say, *writing*. Lévi-Strauss pointed out that all the radical changes traditionally ascribed to the Neolithic Revolution—the domestication of animals, the rise of agriculture, and the establishment of settled villages—took place well before the development, in the fourth century B.C., of what most scholars recognize as writing. "The only phenomenon with which writing has always been concomitant," he asserted, "is the creation of cities and empires, that is the integration of large numbers of individuals into a political system, and their grading into castes or classes." It thus appears "to have favoured the exploitation of human beings rather than their enlightenment."

Condorcet, in other words, had the causality reversed. Writing was a step backward. It had made us less happy, and less free, and perhaps continued to do so. "The primary function of written communication," Lévi-Strauss went on, "is to facilitate slavery. The use of writing for disinterested purposes, and as a source of intellectual and aesthetic pleasure, is a secondary result, and more often than not it may even be turned into a means of strengthening, justifying, or concealing the other."

It is a curious digression, and a funny one, to find three hundred pages into this most literary of anthropological texts. It comes just after an anecdote about the Nambikwara people, whom Lévi-Strauss was studying in the Brazilian Amazon in 1939. He wanted to get an idea of how many had survived the various epidemics that had decimated the indigenous tribes of the Amazon over the previous two decades, so he asked his "friends," as he described them, to take him to their village. Their chief was not pleased with this request: Lévi-Strauss's presence could spark a deadly conflict with other Nambikwara who had very good reasons to dislike Europeans.

If Lévi-Strauss questioned the wisdom of risking his hosts' lives in order to better document their numbers, he did not mention it. Instead he recorded the discomforts of the journey. The group got lost and soon ran out of food: they had been counting on Lévi-Strauss and his colleagues to hunt with their rifles en route, but the anthropologists shot nothing, and were unwilling to share the provisions they had brought. "The next morning, there was widespread discontent, openly directed against the chief who was held responsible for a plan he and I had devised together." The chief and his wife spent the day gathering enough grasshoppers to feed the rest of the group. Lévi-Strauss did not say if he partook, observing dryly that "crushed grasshopper is considered rather poor fare." They plodded on.

Things did not improve when they reached their destination. The chief's "edginess" and the "surly attitudes" of the Nambikwara gathered there "suggested that he had persuaded them to come rather against their will. We did not feel safe. Nor did the Indians." To defuse the tension, Lévi-Strauss distributed paper and pencils, as you might try to pacify unruly toddlers with crayons and colored paper. To his surprise, they all drew the same thing: "wavy, horizontal lines." They were imitating writing, he realized. But the chief, whose

authority was already in jeopardy for having acceded to the anthropologist's requests, took the game one step further. He called the
group together, held up a sheet of paper scribbled with wavy lines,
and "made a show of reading it, pretending to hesitate as he checked
on it the list of objects I was to give in exchange for the presents
offered me: so and so was to have a chopper in exchange for a bow
and arrows, someone else beads in exchange for his necklaces." He
was reminding his people of the wisdom of his leadership, and how
much they stood to gain from sticking with him. "This farce," Lévi-
Strauss wrote, "went on for two hours."

Recounting it years later, the anthropologist's irritation with this
"piece of humbug" had not run dry. He had unwittingly introduced
writing to the Nambikwara and they had taken it up, but not as they
were supposed to. Writing for the chief was not a means of recording knowledge or representing beauty, "but rather of increasing the
authority and prestige of one individual . . . at the expense of others."

It is at this point that Lévi-Strauss, disgusted, begins his dis-
quisition on the unavoidable violence of writing and its origin as a
tool for the maintenance of the repressive hierarchies of early states.
The "extraordinary incident" of the chief's performance of literacy
becomes the occasion for a theoretical reflection on the relationship between writing and political manipulation. His conclusions
are sufficiently bold that it is easy to miss that he is changing the
subject. It was not only the chief, after all, who had attempted to
subdue his restive followers with the sorcery of the written word.
Lévi-Strauss is the one who handed out the pencils and paper, and
it was Lévi-Strauss who brought the crisis on by pressuring the chief
into bringing him somewhere where his presence could cause poten-
tially disastrous trouble. And Lévi-Strauss, in a final, unanswerable
display of mastery, wrote the incident into *Tristes Tropiques*.

Lévi-Strauss does not say if he ever followed through and

distributed the items the chief had promised he would give to the Nambikwara. Presumably he did not. Only at the end of the aside, and then only in parentheses, does he mention that the chief's ruse did not work: "After my visit he was abandoned by most of his people." A few lines later we learn that they "went off into a more remote area of the bush to allow themselves a period of respite." The anthropologist's arrival caused the break-up of the group that he was studying. Lévi-Strauss did not reflect on this, nor on the fact that his own text too was *written*, and was itself part of a long line, dating back to Bernal Díaz del Castillo, and to Columbus himself, of European writings about indigenous Americans that were at every step accompanied by conquest and destruction—"leaving descriptions of what we wipe out," per Ursula K. Le Guin.

If his presence, and his writing, caused harm to his ostensible subjects, Lévi-Strauss did not wish to dwell on it. Better to highlight the violence of *all* writing, shrug, and move on.

≈

One of the things I am trying to do here is not ever shrug.

≈

He was gone when I woke up, or at least when I first opened the door and looked outside. It was about seven. The mattress was gone too, though I suppose it's small enough that he could have folded it and taken it with him.

I didn't run today. It's windy. There's dust blowing everywhere and trash skipping through the yard and my allergies have been terrible. The Rhino was up early, tweet-ranting. He called Bashar al-Assad a "Gas Killing Animal" and responded to a warning from a Russian diplomat that not only would any U.S. missiles that endangered Russian soldiers stationed on Syrian bases be shot down, but "the sources from which the missiles were fired"—meaning U.S. planes, warships, or submarines—would also be targeted.

"Get ready Russia," the Rhino tweeted, "because they will be coming, nice and new and 'smart!'"

"Smart."

Sometimes it feels a little late to mourn the written word.

≈

When Sarah Winnemucca's grandfather, the Northern Paiute chief who would be known to history as Truckee, returned in 1848 from fighting in California alongside one John C. Frémont, he brought his people many gifts and told them of "the many battles they had had with the Mexicans, and about their killing so many of the Mexicans, and taking their big city away from them." Stranger and more marvelous than anything else he brought back, or any of his stories, Winnemucca wrote, was a piece of paper, "which he said could talk to him. He took it out and he would talk to it, and talk with it. He said, 'This can talk to all our white brothers, and our white sisters, and their children. Our white brothers are beautiful, and our white sisters are beautiful, and their children are beautiful!'"

Whenever Truckee and his people came across settlers or soldiers, the chief would ride ahead and show them the paper that he called his "white rag friend." He held it to the sky and kissed it, Winnemucca wrote, "as if it was really a person. 'Oh, if I should lose this,' he said, 'we shall all be lost.'"

It didn't always help. In his absence, the settlers still raped and murdered his people, and the diseases they brought nearly wiped out the tribe. But even after "all these things had happened," Winnemucca wrote, "my grandfather still stood up for his white brothers." He died in 1860, a few months after the conclusion of a war against the whites that he could not stop his people from fighting. Sarah Winnemucca was by his side at the end. The old chief closed his eyes, she wrote, and she and her mother and everyone else gathered around him began to cry, thinking he had died, but he opened his eyes once more, and spoke. "Don't throw away my white rag-friend," he said. "Place it on my breast when you bury me."

Then he died.

The Rhino seems to be backing down. This morning he spat out: "Never said when an attack on Syria would take place. Could be very soon or not so soon at all!" So maybe a full-blown world war won't break out this weekend. Which is good, because L. is here with me. I picked her up at the airport last night. Music was blasting all through the arrivals hall and giant LCD screens blared advertisements for Cirque du Soleil and David Copperfield and J. Lo's shows at the casinos and Gordon Ramsay's restaurant. There were banks of slot machines between the baggage carousels and everywhere else there was room for them, plus a Lamborghini that you could rent for $1,999 a day parked beside the elevators.

L., who had never been to Las Vegas before, couldn't stop laughing, as much at the thought of me living here as at any of the particulars.

Saturday morning we drove northeast on Interstate 15 and watched the desert shift from the bajadas and playas of the Mojave to sudden red sandstone buttes and high, flat-topped mesas as we crossed through a corner of Arizona into Utah. We stopped first outside the town of St. George and hiked through red dunes dotted with sage and thin-spined yuccas until almost by chance we found what we were looking for—a narrow slot canyon covered in petroglyphs. Its walls were straight and dark, maybe twenty feet high and so close that with my backpack on I couldn't turn sideways. A tree grew in the middle of the canyon, its trunk twisted and gnarled. The glyphs were old and faded but, once my eyes adjusted to the dim, startlingly clear: bighorn sheep in profile, wavy lines like rivers or serpents or the surface of a stream, six-fingered hands and circles bisected by vertical bars just like the ones I'd seen in Sloan Canyon, 150 miles away.

We camped that night off a rutted backcountry road somewhere off the Virgin River several miles from nowhere. We set up the tent, lugged the food and water bottles from the car, and went for a walk, climbing down the soft dirt cliffs. Somewhere above us a shot rang out. And then another, and another. High at the top of a butte we could just barely make them out, three silhouetted figures with long-barreled guns, aiming into the setting sun. They kept at it until they had killed it, the crack of each shot rumbling like thunder through the valley in the night.

The next day we climbed a mountain, lay in the shade of the cottonwoods that grew on the banks of the river, and dragged our feet through its cold, fast water. On the drive home we watched the sun set and Venus sink behind it. L., in the passenger seat, tracked the rising stars until the glow of Las Vegas began to dull the sky. Finally the darkness broke and the city slouched beneath us, that golden net tossed over the valley with as much care as a dirty shirt flung on the bed at the end of the day. I got off the freeway just north of downtown. The sidewalks beneath the overpass were crowded on both sides with people camping there in blankets and sleeping bags, their belongings stuffed into carts beside them. It wasn't late, maybe nine o'clock. Only a few of them were sleeping. The rest sat upright on their blankets, staring in silence at the concrete walls on the opposite side of the street.

✻

For how many people today does the very idea of writing provoke first of all anxiety—not the almost erotic thrill of a sentence that shimmers with its own perfect movements, but the dread that meets a certified letter from the bank, a stranger at the door with a clipboard, cops with a warrant, a man in a suit flipping through the pages to find the dotted lines—*signature here and here, initials there, there, there, and there, and one more there*—the memory of gazing down at an impenetrable black maze of letters, your thighs squeezed between the chair and the desk and sticking to both, words swirling, the teacher's voice echoing your name? Writing is love letters too, a postcard from a friend that makes you laugh from far away, sexts from someone you want to sext with, your child's name tattooed above your heart, a brave, forbidden slogan on a wall. Hell, it's poems and novels and it's whatever this is too and it's the all-but-lost art of speaking truth to power but it's also an audit, a foreclosure, a notice to appear. It's the columns in the newspaper that inform you that you're broke because you're lazy and defected, that the rich glow like that because they're virtuous and brilliant and sacrificing another country is not too high a price for rooting out terror wherever it spawns. It's an all-caps tweet from the Rhino. SAD! It's all the bad news of the state.

The anthropologist and political theorist James C. Scott picked up where Lévi-Strauss left off. Writing first appeared, Scott wrote in 2017's *Against the Grain*, at about the same time that the early cities of Mesopotamia were becoming sufficiently complex that we might begin to call them states: "Thousands of cultivators, artisans, traders, and laborers were being, as it were, repurposed as subjects, and, to this end, counted, taxed, conscripted, put to work, and subordinated to a new form of control." All this sorting required greater powers than mortal memory could manage on its own. "It would not be too strong to assert," Scott wrote, "that it is virtually impossible to conceive of even the earliest states without a systematic technology of numerical record keeping," i.e., some form of writing. For its first half millennium, writing remained a form of bookkeeping, a "massive effort through a system of notation to make a society, its manpower, and its production legible to its rulers and temple officials, and to extract grain and labor from it." Only after hundreds of years did anything like literature appear.

In this vision, the appearance of writing, and of civilization itself, functioned primarily to curtail the freedom and happiness of the onetime hunter-gatherers whom Rousseau was not alone in calling *savages*, who until then lived with more leisure, spiritual connectedness, and easy collective joy than we repressed and exhausted urbanites will ever know. (Smohalla, a Wanapum prophet from the Pacific Northwest, exhorted his followers to abandon all the ways of the white man and not to plough or harvest or work in any way, "because men who work cannot dream.") This runs precisely counter to the various narratives of progress that declared Mesopotamia the "cradle" of civilization, placed the invention of writing at the very beginning of history, and traced the stages of its advancement as if it were a child passing through an awkward ideogrammatic adolescence into the full flowering of alphabetic wisdom, which, not incidentally, could reside happily only in the more prosperous districts of the cities of northern and western Europe, and perhaps in a few enlightened households in the northeastern United States.

This counternarrative has tremendous liberatory appeal. If once we were free, perhaps we can be again. If the current arrangement has consigned the species to massive inequality, endless wars of conquest and extermination, and environmental devastation and doom, we are not fated to it. Humans are not ineluctably flawed. We've just made some bad choices (like agriculture), and over the millennia our sensibilities have been muddied, our priorities confused. But if we take this as a judgment on all writing since and conclude, as Scott appears to and Lévi-Strauss quite explicitly did, that writing cannot be severed from its origins as a tool for exploitation, we'll end up with a narrative that shares all the worst flaws of its antagonists. It accepts their rigged definitions—that any writing that does not use written symbols to represent elements of spoken language, and hence fails to sufficiently resemble the system used by Europeans, does not count as "true writing"—and reflects the shape of their arguments in reverse. Rather than inscribing a very particular present, one among many, into the past and projecting it forward, as the inheritors of Condorcet and Hegel did, it carves a single past—the earliest one legible to us—through all of intervening time and reads it, with equal parts satisfaction and dismay, into another particular present. But how much do we know about what we do not know? How many pasts, scratched or traced or smudged in less durable materials than stone or clay, are lost to us? How many do we refuse to see because they fail to align with our desires for the present? How many other presents do we similarly neglect?

What evidence exists includes only those forms of early writing that were sturdy enough to survive the millennia and that archaeologists have been lucky enough to unearth. Scott mentions and quickly skips over the fact that the first evidence of writing found in China had nothing to do with counting reserves of sheep or captured slaves but consisted of questions etched into animal bones and tortoise shells for divinatory purposes. The bones and shells were heated until they cracked so that the cracks could be interpreted, read there as answers from the spirits and the gods, as legible as inscriptions carved by human hands. Scott does not mention any of the clay vessels, tablets, and tiles found in southeast Europe inscribed with glyphs that remained consistent over centuries—the so-called Vinča symbols. Some of them predate the oldest Mesopotamian cuneiform tablets by more than two thousand years, but most scholars dismiss them as "proto-writing." Marija Gimbutas argued that they were fragments of a "true writing system," now lost, an Old European Script "developed to communicate with divine forces." The earliest writing in the Americas—if we discount the many thousands of far more ancient carved and painted petroglyphs—appears on a slab of green stone found in the Mexican state of Veracruz and scratched three thousand years ago with sixty-two Olmec glyphs. Only a few of them depict recognizable objects—ears of corn, torches, an insect or a spider, brass knuckles worn in ceremonial combat—or resemble glyphs used by later cultures. Some archaeologists speculate that the text may represent a "list or some type of register," others that it is "eminently religious in nature." For now we have no idea.

And what of the texts—perhaps even whole systems of writing—that didn't survive? Ceramics inscribed with the Nsibidi script used in the Cross River region of what is now Nigeria and Cameroon have been dated back more than twelve hundred years. As you might expect, considering the tilt of the whole enterprise, scholars of early writing have taken little-to-no notice of this autochthonous African script, but given that Nsibidi symbols are still printed on textiles, carved into wood, traced in the earth, painted and tattooed on human skin, and waved in the air as gestural signs, it should surprise no one if Nsibidi turned out to be far older than that. James Scott quotes the archaeologist C. C. Lamberg-Karlovsky, who suggested that if no evidence of writing appeared in the peripheries of ancient Mesopotamian cities until centuries after it was developed in urban centers, it was not because non-urbanites were any less sophisticated. "Perhaps," Lamberg-Karlovsky wrote, "far from being less intellectually qualified to deal with [the] complexity" of state social structures, "the peripheral peoples were smart enough to avoid its oppressive command structure for at least another 500 years, when it was imposed upon them by military conquest."

Perhaps he's right. Or perhaps writing is larger than we've let it be. Perhaps they used and conceived of writing differently than the inhabitants of early states. Perhaps they drew signs on the smooth inner surfaces of tree bark or etched them in the dirt to convey messages that would last until the rains came and washed them all to mud. Perhaps they wove them on the seams of their scarves and gowns or painted them on walls with pigments that faded in a decade. Perhaps they used them to write about topics more profound or at least more entertaining than the contents of the royal storerooms, outstanding debts, and tributes paid. Perhaps not. Perhaps their brains had not been dulled by domestication—Scott points out that the brains of modern sheep and pigs are respectively 24 and 30 percent smaller than those of their untamed ancestors, and suggests that humans have been domesticated too in the process of taming them—and they saw no need to write it all down because they knew that everything was already speaking, and they could hear it. Perhaps they let their words hang in the air as they spoke them so that they might fall where they might fall, like rain or like stars, or any mortal thing.

Friday we drove across the city and out to Red Rock Canyon and hiked a trail that I had run once before. It was a lot easier to walk it. The trail zigzagged around the backside of a mountain, the vegetation shifting as we climbed from the sparse creosote and yucca of the basins to a near-forest of juniper, oak, and red-branched manzanita. Not much was in bloom, but gauzy white cocoons had appeared on the branches of the desert almond bushes, each one twitching with the caterpillars inside it. I guess it's spring again. Last year in Joshua Tree the cocoons showed up as if overnight sometime in the middle of March, and only in the desert almonds. They didn't like any of the other shrubs. If you looked close you could see the caterpillars writhing inside, crawling over one another to fatten themselves on the tiny leaves. As the weeks passed the cocoons got bigger and the caterpillars grew longer and plumper until one day all of them were gone.

We stayed out long enough to watch the sun set and Venus appear bright in the sky above it just as the light began to fade. On the way home we stopped for fish tacos and saw the news on the TV monitor above the counter: U.S. missile strikes hitting Damascus. Familiar footage of white lights streaking across the sky, the distant yellow glow of an explosion.

≈

By morning, the Rhino was gloating. "Mission Accomplished!" he tweeted. Which of course is what George W. Bush proclaimed fifteen years ago, five and a half weeks into a war that would go on to kill hundreds of thousands of civilians and that has arguably not yet ended. Time is knotted up. If this were fiction it would be too blunt, too obvious. The writers are getting lazy and recycling old jokes.

We left Las Vegas in the afternoon and made it to Joshua Tree by dusk. It's good to be here. A lot is in bloom, far more than in Nevada: the delicate, almost papery orange flowers of the globe mallow, the showy, hot-pink blossoms of the beavertail cactus, the long, white trumpets of datura along the sides of the roads. Here and there you'll find a creosote blooming, its branches bright with tiny yellow flowers. Yesterday the wind howled for most of the day and all of the night, screaming and shrieking and shaking the windows in their frames. It's still singing this morning, making the creosotes dance, rising and falling but never letting up.

&

George and Carobeth Laird settled in Poway, California, in the hills north of San Diego, about fifteen miles from the coast. This was in the twenties and thirties, nearly a century ago. There were no suburbs yet. The area would have been sparsely populated then, rural even, and likely beautiful—low and rolling hills of coastal sage and chaparral, ecosystems that survive now only in scattered parklands. When they drove home from Escondido, the nearest town of any size, George would sometimes pull over to the side of the road in an area then known as Green Valley. There's a subdivision there now: the Westwood neighborhood of the Rancho Bernardo planned community. He would leave Carobeth in the car, she wrote in her final book, *Mirror and Pattern*, scramble up through the thick brush, and disappear awhile.

Archaeologists have since discovered traces of a village there that they believe was inhabited for the better part of the last two thousand years. Most of its remains would have been underground, but George saw the pictographs painted in red on the east-facing sides of a few boulders. He told Carobeth about them. I haven't been there, but I found some photos online. They were different from any petroglyphs I've seen in the Mojave, and entirely abstract: complex, rectilinear labyrinths without clear entry or exit points. They looked like maps to some internal landscape.

What did George Laird do up there? What did he think about as he sat or stood or crouched, staring at those labyrinths? Did he touch the rocks? Did he trace those red lines with his fingers? Did he pray? Grieve? Wonder? Did he figure out how to find his way out of this maze? Carobeth didn't speculate. "I never accompanied him," she wrote, "nor do I recall him asking me to do so."

When she did ask him about the petroglyphs in Chemehuevi territory—the ones in Green Valley were on Kumeyaay turf— George answered only that they were indecipherable, and that they had not been carved by the Chemehuevi themselves but by the "doctors' helpers," the animal familiars who lent their healing powers to Chemehuevi shamans. She did not understand him to be speaking literally: "since the shaman's helpers of the human era were themselves shamans in the story time," his answer was a roundabout way of saying that glyphs had been painted and carved by the tribes' ancestors, a very long time ago. So long ago that Carobeth gave up on guessing what they might mean.

Most scholarly attempts to interpret Southwestern petroglyphs have in one way or another functioned to cast them as meaningless, or at least as so elemental in their attempts at signification that they remain outside of any shared and complex system of signs that might be analogous to language. They are generally understood either as attempts at "hunting magic"—images of prey carved or painted to conjure actual animals—or records of shamanic visions induced through datura and other hallucinogenic plants. If the former, they represent a primitive confusion between signifier and signified, a superstitious belief that the image is the father of the thing. In which case they do not so much produce meaning as document an errant and primitive conception of how meaning works. If instead they were the records of visions, whatever meanings they may have once had resided only in the fevered minds of the shamans who carved them: they could not be understood by anyone who did not share the esoteric terms of the original vision or comprehend their ritual context. But meaning takes shape only in the interaction of complex structures of signification, which are always, necessarily, shared. As

Wittgenstein put it, "A wheel that can be turned though nothing else moves with it is not part of the mechanism." Petroglyphs, by these interpretations, remain far outside the realm of the linguistic, and farther still from the abstract realm of *writing*.

In the 1970s, archaeologists began to speculate that at least some Southwestern rock art may have been tied to astronomical events. Glyphs at several sites appeared to record the appearance in A.D. 1054 of the Crab Supernova, which exploded with such brilliance that it was visible in the daytime and for nearly two years remained brighter than any star in the night sky. A little later, archaeologists sifting through John Peabody Harrington's unpublished notes found links between rituals associated with the solstice and petroglyphs depicting the sun at sites sacred to the Chumash along the coast north of L.A. (Harrington's source on this was an elderly Chumash woman named María Solares, with whom he had photographed Carobeth, shielding her eyes from the sun.) By 1979, scholars had identified ten sites in the southwestern United States where ancient rock art appeared to indicate the location of the rising or setting sun on the solstice. Researchers began camping out on the solstices and equinoxes to observe the play of light and shadow on the rocks and glyphs. Within a few years, dozens more sites had been identified. Whatever else they may have meant and done, at least some of the petroglyphs functioned to mark the cycles of the sun and the stars. They were calendars.

In 1980, Carobeth Laird received a letter from an archaeologist named Beverley Trupe. She and her colleagues had been studying the petroglyphs at a site called Counsel Rocks, in the Providence Mountains, midway between Las Vegas and Joshua Tree—the heart of the Chemehuevi's territory. The glyphs were scattered over a few large boulders at the base of a rocky mesa. Trupe and a colleague named John M. Rafter had spent the winter solstice there and been astonished to see that the long, horizontal lines carved into one boulder pointed precisely to the spot where the sun was rising, and where it would rise only that one day each year. That evening, as the sun set, a splash of light appeared on the concave inner wall of a boulder the researchers had nicknamed Womb Rock. The light swept slowly across the rock, tracing a zigzagged path that had been etched into the stone centuries earlier. At the same time, in another recess in the same boulder, dots of fading light hopped along other lines pecked into the stone. "The lights followed the linear designs as if drawing them with a finger," Rafter later wrote. The petroglyphs had turned the boulders into one giant timepiece.

It was more than that. On the wall of Womb Rock, the archaeologists had found an unusual glyph. It appeared to be an abstract drawing of a vulva—a stock symbol in Mojave rock art—but this one had straight lines radiating out from beneath it. A few inches away was a recognizable sun glyph: rays emanating from a circle. Another carving depicted another vulva, this time between what looked like a pair of outstretched legs, and above them a wavy line, perhaps a mountain range, and above that another sun. For Trupe, it brought to mind a Chemehuevi myth that Carobeth Laird had documented, one that George Laird had told her was "a very ancient telling."

The tale is called "The Sun's Dead Sons," and it is as rich and strange as any of the Chemehuevi stories, a Jungian analyst's dream, replete with rape and matricide, twins and a trickster, hidden infants, divine rage. It begins with a woman who lives alone in a cave in the mountains. Every morning she gets up and climbs to the highest peak to pee, until one morning the sun sneaks up on her and dips his whiskers into her vagina. The woman runs off and before long gives birth to twins. She hides them during the day, but whenever she goes out, the sun comes to visit his sons. He rocks them in their cradles and leaves gifts. When they grow older, they carve flutes out of reeds and, despite the warnings of their mother, delight in playing them loudly. Just as Hunahpu and Xbalanque disturbed the Lords of Xibalba with their ball playing, the twins' music alerts Coyote of their proximity.

Coyote does not dispatch owls but sends the two most beautiful young women from his company, who happen to be the daughters of Owl, to search out the flute players. They soon find the brothers. At night the younger of the two boys—he could only be younger by a matter of minutes, but structurally it seems to matter here—sneaks into bed with Owl's elder daughter. The sisters run home and tell Coyote that they failed and could not find anyone, but it soon becomes obvious that the older girl is pregnant. Coyote knows they lied. If she gives birth to a boy, he announces, he will kill it. Of course it's a boy. His mother does her best to disguise him as a girl, tucking his penis between his legs, tying it down. Eventually, the two brothers show up. They find Coyote's band and sneak into the house where the sisters are sleeping, but the four of them, happy to be reunited, laugh too loudly, and Coyote sends his men to investigate the noise. They find the twins there and kill them both. Owl's daughter, still afraid for her boy, raises him as a girl, but as soon as the child is grown she tells him everything. The boy, now a man, fashions himself a war club and with it kills every member of Coyote's company, including his own mother. Then he goes off in search of his grandfather, but by some twist of solar logic, or just a father's grief, the sun blames the boy for the death of the twins and destroys him with the heat of his rays. "Thus ends the ancient telling." I've skipped a few bits, but that's the basic plot. By the end everyone is slaughtered. Everyone but the sun.

The glyphs that Trupe was studying appeared to illustrate some version of this myth, or perhaps its ancient core—a vulva penetrated by the whiskers, or rays, of the sun. Laird speculated that other characters may have had astronomical referents too: the sons of the sun were the stars Castor and Pollux, which the Romans and many other cultures also associated with twins. Coyote's band formed the Pleiades—a clear image of that star cluster, which in another myth is associated with Coyote's family, was carved into a nearby rock. The vengeful grandson of ambiguous gender may have been Venus, which sometimes acts as the morning and sometimes the evening star. Owl's daughters surely had celestial analogues as well, perhaps the two dim stars that accompany Castor and Pollux. "The story," Laird suggested, "dramatizes the march of the stars and planets across the night sky and their final extinction in the light of the morning." All of them die, just as "all stars fade from the brightening sky until only the Morning Star remains," and it too is shortly killed by daylight.

There's one other myth George Laird told Carobeth that may be relevant. It's also about two brothers with an unnamed single mother, perhaps the same boys. Exhausted and annoyed with them, their mother shoos them off to Snake's house and tells them to ask him for a story. Snake gives them a riddle instead: "Under your mother, dark pulsates, pulsates." The brothers don't know what it means, but they know it's "something bad," so they run home to tell their mom. George explained to Carobeth that Snake's riddle was difficult to translate but referred to female orgasm as well as to actual darkness broken by an intermittent glow. It appears to have been a poetic version of that most ancient and potent of insults: "I fucked your mom."

Furious, the boys' mother rushes over to confront Snake. "When?" she demands to know. "When did you ever make it pulsate darkly?" By way of an answer, he coils around her legs, lies her down, and causes a pointed house, open only from above, to take shape around them. The boys scramble up its side and, peering down through the hole at the top, see their mother and Snake beneath them, enraptured and entwined. They shoot arrows down through the hole, killing Snake and their mother at once.

In mid-spring, John Rafter returned to Counsel Rocks and observed the night sky through a hole in the overhanging rock. It was the same boulder that had been decorated with petroglyphs of the whiskered sun above a woman's open legs, perhaps marking her impregnation by the sun. Through that natural window—like a house with an opening from above—Rafter watched Castor and Pollux, the twins, reach their zenith directly above him. The next day, when the two stars were due to arrive at their lower culmination (the farthest point they would reach beneath the horizon, that is), an arrowhead-shaped shard of sunlight passed through the opening above him and descended into a circle that had been carved into the stone floor, piercing it just as the arrows shot by the twins had pene-trated the intertwined bodies of their mother and Snake. "It was like witnessing a secret unfold before me," Rafter wrote, "a secret that had laid hidden perhaps for centuries."

The idea then is that, whatever else they may have done, at least some Chemehuevi myths, which had roots more ancient than the people we know as Chemehuevi, encoded the motions of celestial objects into narrative, and that the petroglyphs, some of them at least, served both to represent those narratives and to indicate the movement of the planets and the stars. At the same time that the glyphs turned the bare desert rocks into calendars and clocks, they told, and even reenacted, the stories through which the Chemehuevi understood their world, and their place within it. Perhaps they're writing, perhaps they're not, but no alphabet that I know of can play so good a trick with such elegance and economy.

≈

Friday morning we left for San Francisco. We drove east through the desert for I don't remember how many hours until we cut through the Tehachapi Mountains, still green from the spring's meager rains, and down into the San Joaquin Valley. We stopped for lunch in Bakersfield and if I am remembering properly and was not hallucinating a wild-eyed blond woman with her face painted blue danced past the windows of the taco shop. A man slouched on the pavement outside asked to use my phone. His arms were crusted with sores. He wanted to call his mother. He had to get a number from her, he explained, so he could get his money from Western Union. "I know it sounds like a hustle but it's not," he said. I let him call. The line was busy anyway.

We drove for nine hours that day, crossing the Bay Bridge at rush hour. We saw my mother, who was visiting San Francisco— the sidewalks around her hotel were crowded for blocks around with bedraggled people sleeping, smoking, passing the hours, the hills outside the city sprayed purple and yellow with lupine and mustard—and on Sunday we drove east to Sacramento to meet my nieces and their mom for lunch and ice cream. That afternoon we headed up into the Sierras, the temperature dropping as we climbed, Douglas firs and sequoias towering all around us. The road curved through the mountains, the American River burbling along just beneath the highway. Finally the high peaks gave way to rolling hills, browner and barer than on the other side. Desert again. We stayed in a motel in an old copper mining town called Yerington. I read later about the arsenic and uranium contamination in the water that the mining company, Anaconda, had never cleaned up. The same company once employed my grandfather and his father and brothers and uncles in the copper mines of Butte, Montana. This is the same grandfather who later became obsessed with the solar system. He was also named Ben. A year or two ago a flock of migrating snow geese landed in the toxic waters that have filled the open pit that Anaconda left behind in Butte. As many as ten thousand of them died.

In the morning we drove to the Walker River Paiute Reservation, where Wovoka is buried and the Ghost Dance was born. I parked outside the tribal administration building and asked the women at the desk how to find the cemetery. They dug around for a while and came up with a color-coded trash pickup map that marked neither our location nor the location of the graveyard, but they said to turn at the water tower, so I did, and there it was, dusty and almost grassless, the grave markers mainly wooden slabs and crosses neatly whitewashed, faded polyester flowers in pink and orange scattered by the wind over the bare dirt of the surrounding fields.

L. found Wovoka's grave and called me over. There was a low metal fence around it and an archway above the gate onto which his name had been welded in rusting capital letters. I could hear birds chattering in the trees and the wind sighing through the leaves and the soft w'hoo-woo-woo-woo of a mourning dove. Geese were honking over the river to the north. Inside the fence were two graves, each one marked with a headstone of sturdy red granite: one for Wovoka and one for Mary Wilson, his wife. The words "Indian Messiah and Prophet" had been carved across Wovoka's, "Wife of Wovoka" across his wife's. He died one month after she did. There were a few pots of yellow artificial flowers on the bare dirt and some white and purple ones in pickle jars, a pine cone that someone must have brought down from the mountains, and four abalone shells filled with coins and crumbling cigarettes, offerings to the man who could light his pipe from the sun, call rain from the clouds, make ice fall from the sky on a hot summer's day.

I had a couple of dimes and a few pennies in my pocket but it didn't feel right to toss them in. I couldn't bring myself to lift the latch on the gate or to step inside the enclosure. I knew I had no right to it.

I wasn't sure even that I had any right to the grief that I couldn't help but feel, grief for the unburied and the buried dead and the land raped and ravaged all these years and all our unending foolishness and greed and the sorrows that it piles upon sorrows. Right or no right, I could feel the grief sitting in my gut and coursing up and down my spine into my shoulders and my jaw as I stood outside the gate and stared in over the steel bars of the fence, watching the mountains swirl around me in the distance as I tried to imagine the paradise that Wovoka had tried to summon, the white men washed away and, without me and my kind here to witness it, the hills and valleys teeming once again with deer and wolf and antelope and grizzly, the ancestors raised and the recent dead too, the sick and the crippled healed, all of them dancing, singing, no longer in longing and in mourning but with joy. I stood so still that two jackrabbits hopped right past me. They noticed me only when they were a few feet away. Panic froze their eyes.

We drove on to Walker Lake, the last shrinking remnant of a freshwater sea that twelve thousand years ago covered this whole corner of the state. It's still shrinking, faster now than ever. It's half as long and nearly two hundred feet shallower than it was in 1880, when settlers began diverting the river to irrigate their farms upstream. With so little water it became saturated with the salts that occur naturally in the soil and it's now so saline that most of the fish that once sustained the Paiute here have died. Red DANGER signs in English and Spanish were planted a few hundred feet from the shore, but not because of the water. The beach was OFF LIMITS, the signs warned, due to the presence of UNEXPLODED MUNITIONS WHICH COULD CAUSE SERIOUS INJURY OR DEATH.

A few miles down the highway we passed the massive army ammunition depot at Hawthorne, two-hundred-plus square miles of oblong concrete bunkers keeping thousands of tons of bullets and bombs ready for the next world war. After more than a century of my people's presence here, most of us still see only two options for the desert: dig it up or blow it up. We drove on for a hundred miles and then another hundred, the sparse sage of the Great Basin giving way at last to the Mojave's familiar creosote, passing here and there through dead or half-dead mining towns. The wind makes quick work of twentieth-century human housing, tumbling walls and caving in roofs, turning drywall to dust, shredding carpet and insulation, scattering them in the rabbitbrush and rocks.

Farther down, we passed Creech Air Force Base, where pilots in padded office chairs control drones half a world away, spying and surveilling and slaughtering people in Syria, Iraq, Pakistan, Afghanistan, Libya, Yemen, and Somalia, people redefined as "targets." A gray Reaper drone circled the base like a huge, blind crane, the clumsiness of its flight almost touchingly awkward. It was five o'clock and a shift must have just ended. A long line of pickup trucks was leaving the base through the gate and heading south, toward Las Vegas, and home, just like we were.

᠊᠊ᢓᢋ᠊᠊

At the very beginning of this, I suggested that the *Popol Vuh* may have been the only book that successfully records its own destruction. That's not quite right. The text that we have, inked in parallel columns of K'iche' and Spanish prose by Francisco Ximénez in the early 1700s, was likely copied from an older manuscript that had been transcribed into Roman letters in the mid-1550s and which was itself a transcription of another still-older text painted in Maya glyphs. It is the disappearance of that original book—the actual, preconquest, glyphic *Popul Vuh*—that the much-diminished alphabetic text appears to mourn.

If it was anything like the few Maya codices that escaped de Landa's holocaust, the lost original would have been a different sort of book entirely: a single, long sheet of paper made from treated bark and folded, accordion-style, into individual sheets. Translated glyph by glyph it would have made little narrative sense, but it was not intended for the sort of reading we are accustomed to: a solitary, imaginative act of communing between a reader and an author distant in time and in space. All books are portals, but the original *Popol Vuh* would have had more doors than most. It would have included painted illustrations, ritual schedules, and precise astronomical and calendrical tables. In addition to the stories that it told, it would also have functioned as a sort of divinatory manual, or in the words of its English translator, Dennis Tedlock, "as a complex navigation system for those who wish to see and move beyond the present."

The apparently straightforward events that it narrates—the flight of the messenger owls, Hunahpu and Xbalanque's multiple descents to Xibalba and subsequent returns to the world of the living—would have been understood to describe astronomical cycles as well as mythic happenings. The owls' journeys to and from Xibalba

match the rising and setting of the planet Mercury; the twins' as-
cents and descents track those of Venus. Other characters track the
Maya equivalents of the Big Dipper, the Little Dipper, the Pleiades.
The plot at once accounts for the mythical origin of the universe and
describes the complex motions of the planets and stars in a manner
that would permit a skilled diviner to peer into the future as well
as the past. The original *Popul Vuh* would still have allowed for the
more ordinary transports of narrative, albeit enjoyed in a more collective fashion than our books generally are: its heroes' adventures
and the sweep of its plot would likely have been performed for an
audience. The book that remains to us, Tedlock speculated, may
be less a direct transliteration than a summation of the events that
would have been described over the lengthy course of such a performance. For all its liveliness, the surviving, alphabetic transliteration
of the *Popol Vuh* would then be a shadow of a shadow, a flimsy, one-
dimensional substitute for the multidimensional original.

In the version that we have, after interrupting its tale of the
creation to recount Hunahpu and Xbalanque's adventures in Xibalba, the *Popol Vuh* returns to the efforts of the gods to craft a
being capable of praising them. This is why they made us: so that
someone would speak their names aloud, praise them and pray to
them, keep track of the days and the cycles of the heavens. They
failed three times before they succeeded. First they made the animals, who cannot speak, but only moan and cry. They tried making
men out of mud, but their creations crumbled and dissolved. They
tried wood too, but the hearts and minds of the men they carved
were empty and could retain no memory of their creators. Finally
they made men out of corn. These ones were perfect. Too perfect.
They could see through mountains and through oceans. They saw
and understood everything perfectly and without obstruction. Just
like the gods did. This made the gods anxious, so they dimmed our

sight and dulled our understanding, as you might blow into a mirror, clouding its surface. Since then we have not been able to see very far, not without help.

In its final pages, the surviving text of the *Popol Vuh* suggests that the perfect knowledge our ancestors enjoyed had been partially restored. Or at least that a tool had existed that did the trick. Those who had been trained to use it properly "knew whether war would occur," and "whether there would be death, or whether there would be famine, or whether quarrels would occur, they knew it for certain, since there was a place to see it, there was a book." The book was the *Popol Vuh*. We don't know what happened to it, whether it was found and burned or hidden and lost, we just know it's gone now, and our sight as dull as ever.

The paloverdes in the bank parking lot are blooming bright and yellow now. Great-tailed grackles are courting in the trees, their feathers a black so deep and oily that it seems to reflect every color at once. Four Palestinians were killed in Gaza today, one of them a fifteen-year-old boy, shot in the head by a sniper near the fence. A video of the shooting is circulating on Twitter. I didn't watch it. The person who sleeps on the foam mat is a woman, it turns out. I had seen a man on it before, but he seems to be gone. She unrolled the mat in the alley today and sat there, reclining in a narrow strip of shade, feeding the pigeons, smoking cigarettes. It's hot now, over ninety, and it occurred to me while walking to pick up my car from the shop—the air-conditioning chose this week to quit—that Las Vegas might be easier to take if I understood it not as a city but as a dream the desert is dreaming about itself. This makes more sense than any historical, economic, or sociological analysis I have come across so far. It explains more too. Walking in the sun on the dingy, gum-stained sidewalk along the vast, five-lane expanse of Charleston Boulevard, past the EZ Pawn and the Speedee Cash and the empty, potholed parking lots and the immigration and divorce attorneys' offices in low, beige stucco houses, I found the thought reassuring even if it means, as it must, that I am a part of the dream too, and it will soon transmute into another dream, as dreams always do, and then disappear and fade quickly from the recollections of the dreamer, whoever that may be.

2

I meant to get to Jacob Boehme earlier, back when I first read him, after I stumbled across his name in that Rexroth poem in S. and D.'s living room. I've been reading his books and books about his books and his life and it all took a few weeks to digest. Boehme was twenty-five when the vision hit him. The year was 1600 and all was well, or should have been. Boehme had only months before opened his own shop in the town of Görlitz, on what is now the German side of the border with Poland, a few miles north of the Czech Republic. (None of those states existed then.) Born to a peasant family and apprenticed to a cobbler at fourteen, he had ascended to the rank of *burgher* with all its attendant privileges, and his wife, a butcher's daughter, had just given birth to a son. Boehme had done well for a boy from the country, but he had nonetheless fallen into a deep, extended melancholy. In all things he had noticed evil fouling whatever good there was. God did not appear to intervene, for "it went as well in this world with the wicked as with the virtuous." Boehme was "perplexed," he wrote, "and exceedingly troubled," and even in those tempestuous times—the Protestant Reformation was not yet a century old, and myriad strands of belief and dissent swirled through Görlitz and its surroundings—he could find no scripture or creed that gave him comfort.

When it came it came without warning. A glint of sunlight in a pewter dish caught Boehme's attention. His world shattered: "In this light my spirit suddenly saw through all, and in and by all the creatures, even in herbs and grass it knew God, who he is, and how he is, and what his will is." Everything was suffused with love. Boehme felt like a man resurrected, he wrote later, like he had been dead before and only now "perceived and recognized the Being of all beings, the Byss and the Abyss." He could not adequately express what he had experienced, he wrote, "either in speaking or writing, neither can it be compared to anything," but it gave him the mission that would be his sustenance and his torment for the remainder of his life: "to describe the being of God." He had little formal education, but he would nonetheless attempt to get it down in writing, "though I should irritate or enrage the whole world, the devil, and all the gates of hell."

His first effort, the book that would be known as *Aurora*, took him twelve years to write. He lent the still-unfinished manuscript to a sympathetic local nobleman who was sufficiently impressed that he had the work copied and distributed. One copy ended up in the hands of the chief pastor of Görlitz, who denounced Boehme from the pulpit, demanding that the civil authorities punish the shoemaker for his heretical views. (Boehme had been imprudent enough to suggest that God is everywhere, and heaven too, "even in that very place where thou standest and goest.") The manuscript—that copy, at least—was confiscated. Boehme, after a brief imprisonment, was obliged to vow that he would write no more. For six years he kept his word, but when he started writing again in 1618, he could not stop. Between that year and 1620 he completed no fewer than seven works, a task surely not made easier by the fact that he had given up his shop and turned to buying and selling yarn, gloves, books, and other goods, traveling from town to town in a landscape rendered perilous by the outbreak of the Thirty Years' War.

In 1621, Boehme wrote *The Signature of All Things*, which Kenneth Rexroth would read some 320 years later, lying in the shade beside a waterfall, and which I would read another three-quarters of a century after that, in bed in my apartment and hunched over a table in the fourth-floor reading room of the library at UNLV, my hands green in the fluorescent light. The full title is impressive: *The Signature of All Things Shewing the Sign and Signification of the Several Forms and Shapes in the Creation and What the Beginning, Ruin, and Cure of Every Thing Is. It Proceeds out of Eternity into Time, and Again out of Time into Eternity and Comprizes All Mysteries.* Ahem. Like all of Boehme's works, it can be difficult going. By 1621 he had dispensed with the apologies for the modesty of his learning that appear throughout *Aurora* ("Or dost thou think my writing is too earthly?"), but his prose had not grown easier to decipher. Boehme did not develop arguments discursively so much as layer on metaphors and motifs, trying on new ones and circling back to old ones. In *The Signature* he employs the spiraling paradoxes common to mystical theologies, freely combining the technical terms of Paracelsian alchemy with words of his own coinage. The result is often not easily comprehensible: "Understand it thus; *Sul* is in the first Principle the Free-Will, or the Lubet in the Nothing to Something, it is in the Liberty without Nature; *Phur* is the Desire of the free Lubet, and makes in itself, in the *Phur, viz.* in the Desire, an Essence . . ."

Despite the myth that surrounds him, which he did his part to encourage, Boehme was not a barely literate shoemaker informed solely by solitary ecstasies. He read a lot. Through friends and correspondents he had become familiar with aspects of Kabbalah, the Christian mystical tradition, and the Neoplatonist and Hermetic strands in Renaissance theology and thought. Görlitz was home to a significant community of followers of the sixteenth-century Swiss alchemist, physician, and astrologer Paracelsus, on whom Boehme's otherwise mystifying talk of Sul and Phur can be blamed: for Paracelsus, sulphur was one of three basic substances that comprise all things, the others being mercury and salt. (Splitting the word into its component syllables appears to have been Boehme's innovation.) *The Signature* gets its title from the old medieval "doctrine of signatures," or Parcelsus's version of it, which held that all things from the heavenly bodies to the leaves of the smallest plant are related by complex webs of correspondence, analogy, and influence that can be decoded through their "signatures," stamps of invisible essence that show themselves in external appearances. "By the outward shapes and qualities of things," Paracelsus had written, "we may know their inward virtues, which God hath put in them for the use of man."

At the crudest and most practical level this meant, for instance, that herbs with heart-shaped leaves could be used to treat coronary troubles, but Boehme extended the doctrine well beyond the medical and botanical spheres. The signature, for him, was the mark of the divine that gives things in their teeming multiplicity a specific form and character. It is what makes things what they are while relating them to all others, and hence stands in for the sheer, creative exuberance of the spirit. "The greatest Understanding lies in the Signature," Boehme wrote, "wherein Man . . . may not only learn to know himself, but therein also he may learn to know the Essence of all Essences; for by the external Form of all Creatures, by their Instigation, Inclination and Desire, also by their Sound, Voice and Speech which they utter, the hidden Spirit is known; for Nature has given to every Thing its Language . . . Every Thing speaks . . . and continually manifests, declares, and sets forth itself."

This was Boehme's radical and borderline heretical claim, that the signature of God can be read in all things. All of existence—good and evil, pure and impure—is the language of God. Everything is speaking—shouting, even, if you know how to listen. Literacy, learning, elevation to the clergy or to high academic post won't help. Nature forms a second set of scriptures. It too is a book authored by God. Which means that everything is *writing*: "The whole outward visible World with all its Being is a Signature, or Figure of the inward Spiritual World."

There is nothing fixed in this vision, and nothing fated. "There is a continual Combat in the Earth," Boehme wrote. The chain of microcosmic relations is infinite. Everything contains everything else and holds its opposite within it, which means that everything is *writhing*. "In the heavenly Being there is also a Property of a Seething," Boehme wrote at one point; at another that "the Being of Beings is a wrestling Power." God's love and anger are of a single piece. Apparent opposites are locked in constant battle, spawning one another as they spar. A man inclined to goodness might be turned by wrath to evil and an evil man made good with love. Just as for Paracelsus a cure existed for all possible ailments, for Boehme, in his profound alchemical optimism, the same divine principles that populate the heavens exist in full within every individual. If one element is overabundant and causing harm, it can be countered by its antagonists. No ailment exists that cannot be cured.

Three years after Boehme wrote *The Signature*, his old adversary, Görlitz's chief pastor, got his hands on another one of Boehme's books. He again denounced Boehme as a heretic and a "common Disturber of the Peace." He seemed at least as offended by Boehme's departure from his assigned station in life as by the contents of his works, which, the pastor fumed, stank *"abominably* of Shoemakers Pitch *and* Blacking." Boehme was summoned once more before the town council. This time he was encouraged to "take his departure as soon as possible." That was in April. Boehme fled, but returned to Görlitz in November, gravely ill, and died soon after. The local clergy refused to bury him until, after three days, they were ordered to do so by the town council. The pastor to whom the task of delivering the sermon fell prefaced his remarks by explaining that he was there only because he had been ordered to. He would rather be 120 miles away, he insisted, "had such an Excursion been allowed me." The cross that Boehme's followers erected on his grave would be "bespattered with filth, mangled and mutilated," and destroyed before the year was out. It was made of black wood and painted with esoteric symbols—a child resting his head on a skull, a lion in a golden crown clutching a flaming sword and a burning heart, and an eagle with a branch of lilies in its beak, treading on a snake.

❧

More new studies. The Gulf Stream is weakening, and fast. The Greenland Ice Sheet is melting more rapidly than at any time since 1550, which is as early as the data goes. The polar seas are growing warmer, causing the Atlantic Ocean currents that push warm water north from the tropics and cool water south from the pole to lose force. They are weaker now than they have been at any point in the last sixteen hundred years, as far back as climate scientists have analyzed. The consequences of a more complete collapse—which until very recently scientists thought was centuries off—would be catastrophic. During the last ice age, a weakening of Atlantic currents caused winter temperatures to drop as much as ten degrees centigrade, or eighteen degrees Fahrenheit, over just three years. There is no bright side: summer temperatures will continue to rise.

❧

Boehme's influence would nonetheless be great. We know that Leibniz read him, and that Isaac Newton copied out long sections of Boehme's prose by hand. Newton's near deification of "Absolute Time," which "of itself, and from its own nature flows equably without regard to anything external," in this context begins to make a different kind of sense. William Blake, who wrote of holding "infinity in the palm of your hand / And eternity in an hour," admired and adored Boehme. Thomas De Quincey gave a complete four-volume edition of Boehme's works as a gift to his friend Samuel Taylor Coleridge, who had already been reading Boehme for years. In the first decade of the nineteenth century Boehme became the darling of the German Romantics, and it was likely through them, in the university town of Jena, that Hegel encountered the work of the man he called "the German cobbler of Lusatia, of whom we have no reason to be ashamed."

Such ambivalent praise would mark all Hegel's estimations of Boehme. On the one hand, he gave him pride of place as "the first German philosopher" and devoted to him what amounts to a full chapter of his *Lectures on the History of Philosophy,* in which he described reading Boehme as "a wondrous experience," and praised his "profound and German soul." On the other, he expressed near-constant frustration at the homey inexactness of Boehme's argumentation, complaining of a "most frightful and painful struggle between his mind and consciousness and his powers of expression." He at one point referred to Boehme as a "complete barbarian" and elsewhere regretted his "crude and barbarous presentation," condemning the "barbaric fashion" of his speech and the "unmistakably barbarous" quality of his articulation. The usual hierarchies were in effect. Boehme should have counted himself lucky not to have been esteemed a savage. Hegel's main gripe was that, "in order to put his thoughts into words," Boehme had relied on "powerful, sensuous images." Just as hieroglyphs and other forms of ideogrammatic writing still rely on pictorial symbols, expressing the primitive barbarism of their creators, Boehme had failed to ascend to the ethereal abstractions of truly philosophical thought.

For all that, Hegel found much to appreciate. Boehme's main concern, in Hegel's reading, was to reconcile the absolute unity of God with the existence of negativity and evil. If God was perfect, why was the world so royally fucked? Boehme's solution—an expansive notion of the divine that is eternally wrestling with itself, creating both itself and the universe through that constant, teeming conflict—was not altogether different from Hegel's understanding of a human world shaped by God's ceaseless effort to reach a more complete self-awareness, an effort that takes place dialectically, through the interaction of opposing forces. I don't mean to suggest that Hegel got it all from Boehme. Or any of it in as straightforward a manner as "influence." It would be impossible to parse out the depth of Boehme's impact on Hegel, when and how it arrived, where precisely in his thought it fell. It would also be boring beyond belief. But there is nonetheless a clear kinship to which Hegel, despite his distaste for Boehme's not-quite-civilized modes of expression, was not ashamed to confess.

What I want to suggest here is only that it is possible to see something of Boehme's profound, metaphysical optimism flowing through Hegel's understanding of history as Spirit's ascent to self-consciousness via the apparent chaos of human endeavor. In other words, that progress, and Hegel's notion that humanity was advancing through identifiable stages toward a certain goal, was not solely a murderous and narcissistic, Eurocentric chauvinism, though it certainly was that. It was also marked by a faith—worked out via a wide and largely subterranean network of mystical, Kabbalistic, Hermetic, gnostic, and other creosotal influences stretching back over millennia and across continental and religious boundaries— that humans might approach ever closer to the divine because God is seething forth in everything. And somehow this notion, at once ancient and obvious, that everything partakes in the sacred and everything is speaking because everything is alive, that the past and the future both hum within the present because everything holds everything else inside it, would be transmuted over centuries and under the influence of various competing and contradictory forces into another creed entirely. One that insisted, and still insists, that only humans count, and only certain fair-fleshed humans at that, and that with the possible exception of God, everything in the cosmos is dead, or if not actually dead then silent, inert, barren, and empty of intelligence or consciousness, worth only whatever passing use we can put it to before we toss it away. Or as Hegel put it in a lengthy passage devoted to denying the humanity of Africans, "As soon as man emerges as a human being, he stands in opposition to nature, and it is this alone which makes him a human being."

The being of beings is a wrestling power. We can't say that we weren't warned.

✦

After dinner L. and I walked down to Fremont Street. The sidewalks were deserted as always, but close to the old casino district they grew more populated, at first just with ranters and drunks but more and more with shoals of tourists. As the density of bars and the brightness of the neon lights increased we began to encounter wandering herds of white men in pressed shorts and bright T-shirts printed with the names of other tourist destinations, middle-aged couples looking at everything but each other, gaggles of women in their forties already rowdy before ten o'clock. We walked beneath a neon cowboy and a neon martini glass and a neon shoe and then it happened. Fremont Street became the Fremont Street Experience.

To explain that the Fremont Street Experience is a pedestrian mall would be a bit like saying that peyote is a small and spineless cactus, or that Las Vegas is a medium-sized city in the American Southwest. The first thing you see—though you won't likely notice them because your eyes will be on the flashing lights, and the people, and the lights and the people, and because they've been painted in outrageous colors to better blend with the environment—are the bollards and concrete barriers that block the entrance to speeding vans and would-be car bombs. Just past them, beneath the restaurant that offers free quadruple bacon cheeseburgers to customers weighing in at 350 pounds and up, the crowd begins. And the lights, and the music, and the zipline overhead. Above it all hangs a sort of shimmering canopy, a vaulted ceiling that is also a screen, a four-block-long LCD display, bright as Vegas daylight.

I tried not to look up but every time I did it was blaring advertisements for itself, for the experience in which we were already partaking. Every few minutes a new batch of zipliners slid silently beneath it, vaguely human silhouettes blurring by overhead and disappearing into the blinding light above. The crowd—and we were part of it—milled aimlessly, circulating like shards of plastic caught in the slow but inevitable drag of a riptide. They fell in and out of the casinos and bars and paused for selfies and chatted and even flirted here and there but mainly they let the current take them, their gazes sliding covertly from side to side as if they were searching for something shameful. Almost no one smiled. Violence—muted and disguised but tangible still—flowed in the spaces between bodies. L. asked me if she was imagining it or if I felt it too. I reassured her that I could feel it buzzing past.

Cover bands were blasting country rock and heavy metal from competing stages at opposite ends of the thoroughfare. Two angry magicians and a crew of tired break-dancers worked the crowd. Other performers had grifts that were harder to pin down. There was a young woman in boxing gloves—for a few bucks you could take a photo of her punching your spouse—and there were pretty girls in short shorts and bikini tops waving handfuls of bills in the air. The hustle was universal. There were two young black women in shiny leather bras with whips tucked beneath their arms, baby doms waiting for a heavy-pocketed sub to stumble past. There was an obese goth with her top off, white breasts spilling over her belly, her nipples Xed over with duct tape. Her partner slumped beside her, sullen in a grungy bear suit. There was a muscle man in a mask and platform heels, a topless woman wearing most of a dirt-smeared nun's habit, a wheelchair with miniature American flags protruding from the spokes and armrests, its aged occupant nodding off over a Sharpied sign that said he was a veteran. Even the guy lugging the WELCOME TO LAS VEGAS NOW GET THE FUCK OUT sign was taking tips.

We dodged into the Golden Nugget for a drink, but the people inside were too grimly purposeful, as if they were all being tugged by invisible threads to the tables and barstools and slot machines, and there were security cameras everywhere and the gambler closest to us had a shaved head and an iron cross tattooed on his triceps and even the relative quiet of the bar was vibrating with despair, so we gave ourselves up to the Experience again and ejected ourselves onto the first open side street. A block away the streets were dark again and almost quiet. We gulped it down. The emptiness that had before seemed menacing now felt like a salve. An old, white-bearded man in a wheelchair looked up as we walked by. One of his eyes was clouded and dead. I reached into my pocket for a dollar, but he didn't want money He just asked me for the time.

Borges wrote about Boehme as well. He mentions him, anyway, in a story called "The Secret Miracle." His protagonist is one Jaromir Hladik, a Czech writer imprisoned by the Gestapo and awaiting execution. (Borges had a thing for prison cells.) Hladik is the author, Borges writes, of a history of the diverse notions of eternity so far devised by men, another work on the "indirect Jewish sources of Jakob Boehme," and an unfinished drama in verse. The story contains its share of Borgesian folds and doublings, dreams and fictions within the fiction, hallucinations and dreams within those, but Boehme gets no further elaboration, at least not explicitly. He is present enough, though, in the story's suggestion that being teems with hidden abundance, and in a notion of writing that requires no ink or stylus.

Hladik, terrified, passes the hours of his imprisonment imagining the circumstances of his execution in exhaustive detail, then varying them slightly on the theory that reality rarely overlaps with our expectations: if he can run through all the possibilities, perhaps he won't be shot. Finally he despairs. He begs God for just one more year to finish writing his play, so that he will have accomplished something and won't have lived in vain. He dreams that night of the Clementinum, the great Baroque library in Prague. A librarian tells him that "God is in one of the letters on one of the pages of the Clementinum's four hundred thousand tomes" and then takes off his glasses to reveal two dead eyes: he has gone blind searching for it. At that moment another patron approaches the desk to return an atlas. Hladik flips it open to a map of India and touches one of the tiniest letters. A voice speaks suddenly from everywhere and nowhere: "The time for your work has been granted."

Hladik wakes. Two soldiers open the door to his cell, march him to a courtyard, and offer him a final cigarette. Their sergeant gives the order to shoot, but the firing squad does not fire and Hladik doesn't die. Time stops. Everything is frozen: the wind, the sergeant's arm, the shadow of a bee. Eventually Hladik understands. The year he has been granted will transpire within this frozen moment, "between the order and its execution." He can finish writing the play, but it will exist only in his mind, and in his memory, in a year folded inside an instant: "He did not work for posterity nor even for God, whose literary preferences he knew little of. Meticulous, still, and in secret, he wove his high, invisible labyrinth in time."

≋

L. and I both needed to replace the batteries in our phones so she made us an appointment at the Apple Store, which happens to be in a mall called the Forum, which is inside a casino, Caesars Palace, on the Strip. I protested but quickly gave in. After four months here I was curious. We drove down on Sammy Davis Jr. Drive, past the strip clubs and massage parlors and the shooting range where for $2,500 you can crunch a car beneath the treads of an M1A1 Abrams tank. The gleaming, gold slab of the Rhino's hotel towered over all of it.

It was easy to get lost between the parking structure and the mall and for a while we were stuck in a strange elevator loop that I didn't think we would escape from but when we finally walked past the last bank of slot machines and through the marble arch that opened onto the tiled floors of the Forum I was too stunned to speak. I barely noticed the colonnades of pink and white marble; the enormous bare-breasted caryatids; the giant, cloud-puffed murals; the painted figures in fake windows opening onto mock terraces above the faux streets through which we plodded, gawking, lugging along bodies that, in all their gratuitous realness, felt suddenly grotesque.

I couldn't take my eyes off the ceiling. By which I mean the sky magically projected onto it. I had seen photos online, but they had not prepared me for the uncanny sense of spaciousness and depth, the softness of the shifting light and the transmutations of the clouds and the colors, blues sliding slowly across purple into pink edged with gold, closer to a dream of a fantasy of an aurora than to the banal splendors of any actual sunset. Every time I looked up it had changed. The clouds took on new shapes. Day fell into twilight, dawn into day, transfiguring the shade and texture of the light and lending something like grace to the dumb objects in the windows of the shops—the pointy shoes and lacy underpants, handbags and sneakers, sunglasses, watches, suits.

We ducked into the Gap and I felt instantly depressed by the blandness of the light and the unchanging sameness of the garments. When we stepped out again the rotunda outside the Cheesecake Factory had been radically transformed. The screens above the colonnade, which earlier had been displaying gigantic images of fish, as if we were cockroaches gazing up at an aquarium, were bright instead with flames. More flames—I could feel their heat—rose from hidden jets around the fountain at the center of the rotunda. An animatronic sculpture of a dragon was folding its wings and descending slowly out of sight as smoke or steam rose around it.

People had their phones in the air. It was a show, *The Fall of Atlantis*, repeated every hour on the hour, always the same. We had just caught the end of it, but our phones would not be ready for a while, so I went back later to watch from the beginning. The narrative was simple and, given the fuzzy acoustics, incomprehensible, but I watched it again later on YouTube and figured it out. The actors, crudely mechanized mannequins that rose up on hidden lifts from somewhere beneath the fountain, told the story of a kingdom "destroyed by foolish pride" and torn apart by the greed and ambition of its rulers. The aged king of Atlantis had to choose a successor. His daughter, blond and busty, ruled over water. His son, armored and gripping a flame-spitting sword, presided over fire. They fought. The screens displayed tumbling waves lit orange with flames. One element then the other appeared to triumph. The seas are rising, fires spreading everywhere. Fire won. The screens went gray and then returned to their usual loop, advertising gift cards. "Shop," they read, "as the Romans do."

≋

As always, there are wheels within wheels. Recall the vision beheld by the Maya priest in Borges's "The Writing of the God" of a wheel that was not in front of or behind or beside him, but that was "every-where at the same time." Perhaps Borges had read *Aurora*, in which Boehme describes God as a wheel comprised of seven wheels, each at once creating the others and inside the others, each angled in a different direction and all of them turning at once. It could travel in any direction, Boehme wrote, without ever pausing or turning, and the more a man "beholdeth the wheel, the more he learneth its form or frame; and the more he learneth, the greater longing he hath to the wheel; for he continually seeth somewhat that is more and more wonderful."

Boehme wrote a lot more about this wondrous "alwaysone wheel" and the seven wheels within it and their correspondence to the Seven Spirits of God and to the stars and the "planetic wheel"— by the contemporary count there were seven planets, the sun and moon plus Mercury, Venus, Mars, Jupiter, and Saturn. At times he seemed to suggest that God and the stars, "the wild rough stars," are one and the same ("if we consider rightly of the sun and stars, with their corpus or body, operations and qualities, then the very divine being may be found therein"). Elsewhere he took pains to insist that they are not, and to warn that you should not make the mistake of worshiping the stars or praying to them. Still elsewhere he confessed that he himself might have sometimes been in error, having caught only "a glimpse of the great God" because "the wheel of nature whirled about too swiftly, so that man, with his halfdead and dull capacity or apprehension, cannot sufficiently comprehend it."

☙

It occurs to me that all the divisions I've drawn up are false, that I've been foolish, that the forces that drive the desert, that create and created it—the screaming wind, extremes of heat and cold, the ardors of insects and rodents and birds, the slow growth of fungi, sudden floods, the swirl of stars—are the same ones that drive the city of Las Vegas: the voraciousness with which it builds and destroys, its endless thirst, its fear of the dark and of any brief stillness or vacuity, the hungers of its visitors and of the corporations that run the casinos, of the police who serve them and the people who sleep in the streets, the brightness and beauty of its lights. There was nothing to find here, and nothing lost. I never left the desert. This is it.

We went out for a drink last night, to the Stratosphere. It was Friday and we'd been working hard all week. We got lost in the casino, which was no accident, I suppose. It's like IKEA without even the pretense of a path: you are not meant to find your way out. Perhaps the pictographs that George Laird used to contemplate on the road through Green Valley were not labyrinths but maps tracing secret routes from the parking structure to the street in all the casinos on the Strip, gifts delivered via vision, dream, or trance, from the people of the past to whatever we are these days. We watched a young Asian woman play roulette with an impressive display of boredom. With each spin she spread another hundred dollars of chips across the table, half on corners, half on numbers, and lost again. L. fed a dollar into a slot machine, tugged the handle, and watched the numbers twirl.

The elevator operator was young and acned and in a far better humor than I would be if I had to spend eight hours a day smiling at strangers in a vertical casket. We shared the elevator with a couple on their honeymoon. They were big people, from Seattle, they said, and some combination of drunk and stoned that caused them to slump against the walls while speaking with great and expectant solemnity about how weird it is to wear your hair up when you're used to wearing it down. Above the door the floors clicked by until we reached 107. The doors opened. We walked up a short staircase and down a short hallway and there it was, the city extending itself in all directions until it reached the black border of the mountains we knew were there but could not see. Between them and us a great shimmering blanket of light wove itself out of the nothingness, long strands of gold flecked with blue, pink, green, and violet, as beautiful as any sky.

L. gasped. I hadn't warned her. "Somebody just fell past the window," she said.

❧

In the first book he wrote after breaking his six-year silence, a book that took him two years to write, Boehme twice mentions an owl. It's a pretty paragraph. Employing the "powerful, sensuous images" that would so irritate Hegel, he compares the world to a "Bath of Thorns and Thistles" and addresses a hypothetical reader. "Behold," he writes, "thou poor Soul in thy Bath of Thorns, where is thy Home? Art thou at Home in this World?"

The question is rhetorical. He is talking to himself. The answer is obviously no.

Why, he goes on, don't you seek esteem in this world? Why not go after riches and honors and pleasure? Everyone else does. Why not makes things easy for yourself? Boehme was my age when he wrote this, forty-four or forty-five. He had been living for years off his friends' generosity, generally broke and scorned by the authorities, moving from place to place through a landscape torn by war. These questions surely plagued him.

"Why dost thou suffer thyself to be despised and abused by those that are inferior to thee, and know less than thou?" he asks. "Why dost thou make thyself a Fool to the World, and art everyone's Owl and Footstool?"

It's a strange usage. In seventeenth-century Lusatia, apparently, or at least in the verdant environs of Boehme's imagination, the owl did not solely or primarily signify wisdom, misfortune, or death. It was not a messenger or the avatar of a goddess. It was an outcast. It was the bird that every other bird despises, "who one and other will have a Fling and a Pluck at it."

He says it again a few lines later. Why not do like all the others, he asks himself, and make peace with the hypocrites? "Then thou wouldst be beloved, and no Body would abuse thee; and thou wouldst be more safe and secure in thy Body and Goods, than in this Way, wherein thou art but the World's Owl and Fool."

≈

I spent the last few hours, okay, the last few days, fretting, trying to put together the pieces that have been piling up all week: the Rhino's secretary of defense meeting with his Israeli counterpart and immediately afterward telling Congress that a direct military confrontation between Israel and Iran is "very likely," the Rhino's secretary of state flying to Riyadh to touch base with the crown prince before announcing that "Iran destabilizes the entire region," an Israeli strike on a Syrian base that killed eighteen Iranians, Netanyahu clownishly making the case for war (again) and flying to Moscow tomorrow to meet with Putin, the Rhino expected to pull out—also tomorrow—of Obama's nuclear deal with Iran despite warnings from his own cabinet that such a move would risk world war. It wasn't hard, really, to put it all together, but it's hard to know what to do with it.

On the bright side, a new study published this morning predicted that in five billion years, when the sun at last burns through its stores of hydrogen, it will not quietly snuff itself out and fade into the cosmic darkness, as previously believed, but will stretch and swell and finally explode into a gorgeous, glowing ring of plasma.

Such planetary nebulae, one of the study's authors beamed, "are the prettiest objects in the sky."

Of course, owls are not outcasts or omens or messengers. They're birds. They see the world as birds do, and we don't likely figure much in any of it.

The Chemehuevi called Mt. Charleston *Nivaganti*, which Carobeth Laird translated as "Snow-Having," though "Snowy" would likely do just as well. They used the name to refer to the entire snowcapped range of the Spring Mountains, which overlook Las Vegas from the west. Until Coyote screwed up and got his brother killed and they ran off together to the north, bringing the Story Time to a decisive close, Coyote and Wolf lived on Nivaganti, in a cave, when the earth was still covered in water. It was from there that Coyote set off after Body Louse in her tiny, flapping, jackrabbit apron, and to there that he dragged the basket filled with all the people of the earth. For the Chemehuevi and the Southern Paiute, Nivaganti was Mount Sinai and Olympus, "the heart of the Storied Land," a place buzzing with numinous energy, the holiest site in a landscape that had been inhabited once by gods.

It is also on the high, exposed slopes of the Spring Mountains, and not many other places on the planet, that bristlecone pines grow. Bristlecones are the oldest non-cloning organisms on earth: mere adolescents next to grandmother creosote, but ancient all the same. It's also a good twenty degrees cooler up there than it is in the asphalt plains of Las Vegas, so on Sunday, when the temperature was flirting with one hundred degrees, we drove up into the hills.

The trail began at over eight thousand feet. From there it was a steady, steep climb of almost an hour before we reached them. The bristlecones were on the leeward side of the mountain, unbent by wind, and didn't look radically different from the other pines and firs that dotted the slopes. Higher up and on the other side, where nothing sheltered them from the violence of the wind, they looked like a different species: thick-trunked but short, with gnarled and twisted branches. Most of the adult trees were probably between one and two thousand years old. One in the White Mountains, to the northwest on the California side of the border, has been dated to more than five thousand years old. Their wood so effectively resists decay that they stay standing for millennia after they've died, until the rocks at last erode from beneath their shallow roots.

And so it was up there: the dead trees lay between the living, and stood still tall among them, their trunks smooth and hard and bare of bark. The peak of Mt. Charleston, snow-having into May, was perhaps a mile away. A sheer bluff rose about half that distance to the north. The trees stood, seeking no one's company, their roots and branches splayed as if to welcome the cold air and unforgiving wind. It didn't sound at all like the wind in the desert below: not a howl or a shriek but a gentle rush, like the push of water through a stream. The silence between gusts was absolute.

I found a toppled tree and sat. It felt good to touch the trunk and branches, safe somehow, and calming. L. fell asleep with her head on my knees. I watched the clouds blow by overhead. In the valley a raven cawed. Ants crawled over gray rocks spotted with orange lichens. The sun sank and the pines cast crooked shadows over one another's trunks, mingling like that. Almost all of them would have been standing when the Chemehuevi still gathered here each year. Some were likely already two thousand years old when Rafael Rivera wandered into the basin beneath us to the east, and when the

Mormons built a fort there twenty-six years later. Bristlecones grow so slowly that some of their needles, still green, are older than me. Las Vegas is no more venerable than a twig. It is almost certainly more fragile. I could see it down below, a hazy, geometric scuff on the far desert floor. I could even see the Stratosphere, just barely, like a pin half-buried in the sand.

The other day I was reading a pdf of Frances Yates's *Giordano Bruno and the Hermetic Tradition*, the first two chapters of which summarize the works of Hermes Trismegistus, the mythic figure associated with the Greek god Hermes and the Egyptian god Thoth, to whom the varied writings grouped together as the *Corpus Hermeticum* were for centuries attributed. The god Hermes, whom the Greeks considered identical to Thoth, was like him a mediator between gods and mortals, a guide to the paths that stretch between us and the heavens. Athena may have had the owl, but Hermes/Thoth was the actual messenger, the patron of travelers, thieves, and seekers. In his more ancient form, as Thoth, he was the scribe of the gods—at times depicted with the head of an ibis or the head of a baboon, or with the face of a dog on a baboon's body. It was Thoth who brought hieroglyphs to humanity, and with them astronomy, medicine, botany, mathematics, gifts of all the many forms of knowledge of the cosmos.

His supposed heir, Hermes Trismegistus, or "Thrice Great Hermes," was likewise a sort of gateway between divine wisdom and human ignorance, passing on knowledge that had been lost for generations. In *The City of God*, Augustine wrote that Hermes Trismegistus lived "long before the sages and philosophers of Greece," only two generations after Moses, and that he was a grandson of the god Hermes and hence a cousin of Prometheus, who stole fire from Zeus and presented it as a gift to mankind, just as Thoth had given us writing. Things get a little tangled if you let them: According to Hesiod, in retribution for Prometheus's theft, Zeus dispatched Pandora to mankind and had each god load her with a gift, each gift a curse to be unleashed on mankind. Hermes, the messenger god, grandfather of the thrice great-grandson, contributed untruth: "lies and crafty words," all the deceptions, manipulations, and ambiguities of which language is capable.

The texts attributed to Hermes Trismegistus are a diverse bunch, and not nearly as old as Augustine supposed. They were written in Greek and were likely composed by several different authors living in Egypt in the early centuries of the first millennium. In them theology, cosmology, magic, and the seeds of what we would now call science—astronomy, medicine, botany, mathematics—are inseparably linked and, Frances Yates argued over the course of her career, form a hidden and much-repressed current, a secret lineage that in various ways informed many of the men we recognize as the great minds of the European Renaissance and the Enlightenment: from

Bruno, Ficino, and Pico della Mirandola to Bacon, Leibniz, and Newton. In other words, the proudest treasure of what we persist in calling Western Civilization, its tradition of empirical and critical reason, was until the eighteenth century in fruitful and more or less constant conversation with esoteric works of Egyptian, gnostic mysticism attributed to the grandson of a trickster god.

Don't feel bad if you've never heard of them. In the years that followed, this lineage would be meticulously erased. The strands of thought and belief with which it had fruitfully intersected—Jewish and Christian mystical traditions, Paracelsian alchemy, Neoplatonist philosophy—would be marginalized and mocked. Less than a decade after Newton's death, the German historian Johann Jakob Brucker published a monumental, and monumentally influential, history of philosophy, the first of its kind, in which he characterized the Hermetic texts—as well as Kabbalah and Neoplatonism—as *Philosophia Barbarica*. That word again. They were, to Brucker, corrupt, pagan superstitions that stood opposed to the eminently rational history of Christian, European truth-seeking. Many of the entries in Diderot's *Encyclopedia* would be cribbed straight from Brucker, whose ideas were smuggled hence into the heart of the French Enlightenment.

This was precisely the same historical moment that the grand narrative of progress, that glorious arc of Reason that had traveled to Paris and London from Athens and Rome, was beginning to take shape. Brucker's work had been published in Latin by 1744, the *Encyclopedia* beginning in 1751, Turgot's "A Philosophical Review of the Successive Advances of the Human Mind" in 1750. Time was just then finding a cozy refuge in the waistcoat pockets of the English bourgeoisie, the earliest coal-powered steam engines were being used to pump the groundwater out of flooded mine shafts, and humans were beginning to transfer the earth's ancient deposits of carbon up into the planet's atmosphere. It was also the same moment that, as Martin Bernal documented, the Greeks were being elevated and all traces of Africa erased. Athena scrubbed up nice. She looked great in white marble, and slipped past with little effort. Baboon-faced Thoth didn't stand a chance. For most of the next two centuries, the texts of the Hermetic corpus would be consigned to the forbidden and derided margins, out there with Lilith and other spirits of the night. To take them seriously got you labeled an occultist, a swindler, or a quack. Like a forgotten continent, the Hermetic corpus would not be rediscovered until the middle of the twentieth century, when Italian historians, and later Frances Yates, began to rescue it from the disrepute into which it had been banished for so long.

In the chapters that I was reading, Yates quoted the original texts, or at least her English translations of Marsilio Ficino's Latin translations of the Greek originals, which were themselves in conversation with both the Neoplatonist philosophers and far older traditions of Egyptian wisdom literature. In 1463 Ficino had translated the Greek texts into Latin at the urgent request of Cosimo de' Medici, who had them brought to Florence from the monastery in Macedonia where his agents had discovered them. Cosimo was in a hurry. He already had Ficino working on a translation of the complete works of Plato. Put that aside, he told him, do this one first. Cosimo was old and wanted to read the fabled Hermes Trismegistus before he died. Plato mattered less.

The line that caught me was this one: "Contemplate through me . . . the world, and consider its beauty. See the hierarchy of the seven heavens and their order. See that all things are full of fight." That last idea got me, not only for the delightful inelegance of its expression—five-year-olds are "full of fight," and boastful drunks— but for the notion behind it, which we have seen also in Boehme, that the cosmic order in all its intricate beauty teems nonetheless with strife. (Whether or not he had the opportunity to read the texts directly, Boehme was certainly exposed to Hermetic thought by way of Paracelsus.) But Yates didn't use quotation marks and it wasn't clear if her citations were exact or if she was paraphrasing, so I looked up the original text online. Or at least a modern English translation of what we assume to be the original, which was surely copied and recopied many times before it fell into the hands of Cosimo de' Medici's scouts. Yates was quoting, it turned out, but she left out a lot and elided the ellipses. The scanner that digitized her book had apparently mistaken an *l* for an *f.* "All things are full of light," it should have read. To be sure, I went to the library later and checked the pdf against a hard copy. Yep. Light, it said. Not fight.

Still, *fight* seems more correct. Even in this long chain of transmissions and translations over the centuries from language to language and form to form, words tangle, wrestle, fight. Thoth gave us writing and truth. Hermes gave us misunderstanding and lies. The Greeks understood them to be a single god. I kept searching, reading, scouring the internet, and came across a blog post by a Dutch scholar named Wouter Hanegraaff, in which he revealed that the most widely distributed version of Ficino's translation of the Hermetic text the *Pimander*—the Treviso edition of 1471, which spread throughout Europe and was used as a basis for later editions and other translations—was unauthorized, and hastily and shoddily produced, and introduced so many errors into Ficino's text that much of it was rendered essentially unreadable. What passed for sacred mystery was merely garbled nonsense. If anyone noticed they didn't complain. The text's influence was enormous.

≥

L. flies off tomorrow, which means I have only one week before the fellowship ends and I leave Las Vegas too. I can't pretend I'm sad about that, but I will miss the desert. And of course, I'll miss L. We'll only be apart for a month before I join her again, far from the Mojave, but the usual pre-parting depression has already settled on my shoulders, and the generalized anxiety of being far apart when everything feels so uncertain. On Sunday, the Rhino suggested that if he doesn't get his wall, he may "have to think about closing up the country for a while." L. is not a U.S. citizen. The Rhino did not elaborate or explain what he meant, but he said it again. Like an Alzheimer's patient, or like a man accustomed to the feeling of reality slipping away, he tends to say things twice: "We may have to close up our country to get this straight, because we either have a country or we don't." So true. It's more than likely that he meant nothing by it and had not considered his words before speaking, but there is no way to know which blusterous absurdity will get caught like a burr in the raging vacuity of his mind, or to guess what's coming next.

Last night the sky was clear so after dinner we did what I had been meaning to do for months. We got in the car and headed out of the city with no destination but darkness. I drove north on the interstate, turned off on Highway 93, and kept going, past a gypsum mine and two power plants until Las Vegas was barely a yellow blur on the horizon. I pulled over and turned off the engine. The crickets screamed. When I opened the door the sky made me dizzy. That was the idea. There was no moon and the darkness was sufficiently absolute that it revealed itself as something more than a uniform blackness, something with volume, layers, depth. I laid a towel over the hood and we shimmied up and made ourselves as comfortable as we could while we waited for our eyes to adjust.

In the two weeks since we camped in Utah—the last time we were able to see the stars—Orion had set and Taurus had disappeared beneath the horizon. After half an hour we had sufficiently oriented ourselves that we could turn in place and identify the constellations all the way around from Corvus to Virgo to Hydra to Leo to Canis Minor to Gemini to Auriga to Perseus to Cassiopeia to Cepheus to Polaris and the Little Dipper to Draco and a corner of Cygnus to Hercules to the Corona Borealis to Boötes. The Dipper was almost straight overhead. We could make out the paws of Ursa Major, and its snout. "It looks more like an anteater than a bear," L. observed. "It looks exactly like an anteater."

Eventually my eyes and brain got tired. The tenuous order of the constellations was too much to hold on to and I let the sky blur again to a chaos of pinprick lights, bright and dim, near and far, yellow, red, and blue. A shooting star streaked close to the horizon, and after a few minutes another one. I made my wishes. In the hour we had been sitting there the sky had already shifted. I squeezed L.'s hand, happy to be reminded of the expanses between stars, the vastness of even the dimmest lights above us, to remember that the sliver of the spectrum that our eyes are capable of seeing catches only a small and paltry portion of the energy coursing through the universe, that if we could also see the microwaves and radio waves and gamma waves and infrared and ultraviolet light leaping between and within the galaxies, the dark emptiness of space would seem neither empty nor dark, but teeming.

It was almost eleven when we got back in the car and headed south, to the freeway, and to Las Vegas, oozing brightly across the desert floor. "Like radioactive algae," L. said, which was exactly right. Soon we could make out the Strip and even the Stratosphere, which, from this distance, L. observed, looked just like a middle finger.

≈

Perhaps old triple-great Hermes never really went away. Was not the very idea of progress also dependent on the conviction, shouted throughout the Hermetic corpus, that heaven is here for the taking and that humans, despite our mortal bodies, are in essence godly, capable of understanding all that exists, and even of sharing in the creative powers of the divine? "Unless you make yourself equal to God, you cannot understand God," the Hermetic texts advised, and then explained just how to do it: "Make yourself grow to a greatness beyond measure, by a bound free yourself from the body; raise yourself above all time, become Eternity; then you will understand God." If that wasn't easy enough, the text went on: "Believe that nothing is impossible for you, think yourself immortal and capable of understanding all, all arts, all sciences, the nature of every living being."

This, even long after Hermes's banishment, would remain the creed underlying so much of modern scientific endeavor, that all things could be known and all of nature mastered, that man, or at least some civilized subset of mankind, was marching on a path toward perfection. He—and until very recently, it was always only *he*—could understand the behavior of the farthest stars, the nuclei of cells and the viscera of atoms, the nature of tortoises and finches, and also of the unfortunate savages who could not walk this path themselves.

Could it be, then, that we owe our understanding of time— dull time, that we wear like a leash, the time that's always running out, that drags us toward the grave and that we yet never have enough of, empty, fragmented, insulting, oppressive, insufferable time, blinking away on our iPhones, measurable on a management consultant's spreadsheet—not just to ancient wounds and the demands of capital and conquest, but also to the undimmed ecstasies of two-thousand-year-old Egyptian gnostics convinced that "Eternity is the Power of God," that nothing ever begins or ends, and

that, in Yates's paraphrasing, "in this divine and living world, nothing can die and everything moves"? Can even time in all its vastness
contain such contradictions? Where, in the substance of a moment,
would they fit?

≈

The Rhino withdrew from the Iran deal. Time is spiraling again.
One hour after he announced that the United States would no longer honor the deal Obama and the EU had struck with Iran, Syrian
media reported that four Israeli missiles had been fired at an Iranian
military compound on a Syrian army base south of Damascus, killing more than a dozen soldiers. At least half of them were members
of Iran's Revolutionary Guard.

All of this, and L.'s departure, and my own next week, was on
my mind last night. I was already sleeping poorly, or not at all, when
the woman on the foam mat began shouting. She continued on and
off all night, arguing bitterly with some absent interlocutor, cursing
him or her or all of us, falling silent for a while then picking it up
again.

≈

Perhaps we can now begin to make sense of Walter Benjamin's
conviction that the past contains within it an orientation toward
its own redemption, that what he called the "time of now" is "shot
through with chips of Messianic time." (Kenneth Rexroth used the
same word: "scattered chips / Of pale cold light that was alive.") If
eternity, all the past and every future, flits through every moment,
then we can grab it there.

Right here, in other words.

For Benjamin this could only be a political act. It meant

rejecting any structures that relied on the exploitation of labor—which is to say not only our muscles and our skills but our *time*, being as it pulses through our veins—or on the "mastery of nature," which was, he suggested, of a piece with the exploitation of human beings. And it meant rejecting the ideology of progress, the slumberous fantasy that history will carry us to some better land. It would not. Seen without the gilded lies that comprise what we call "civilization," history is an assemblage of massacres, mass enslavements, conflagrations, a growing accretion of ruins. Time had to be "blasted out" and history blasted open. Only then could it be redeemed, and with it us.

This was not, for Benjamin, a choice. It wasn't that redemption lay behind some distant gate at the end of a path that we could choose not to walk, that there were other, smoother and easier roads that we might take with less effort, on which we might nonetheless survive. Then, as now, the only other way led to extinction. Benjamin wrote the "Theses on the Philosophy of History" early in 1940, in the months following his release from a French internment camp in an empty château near the city of Nevers. The war had already begun.

In June of that year, German troops entered Paris. Benjamin fled first to Limoges and then to Marseilles. He left a handwritten copy of the "Theses" there with his friend Hannah Arendt, another German Jewish philosopher who was, like him, hoping to secure passage to the United States by way of Portugal. That September, Benjamin made it as far as the small seaside town of Portbou, on the Spanish side of the border with France. Informed by the local police that he would be delivered the next morning to the French authorities—and thence, almost certainly, to the Gestapo—Benjamin, ill and exhausted, despaired. Alone in his hotel room, he swallowed an overdose of morphine tablets. His body still lies in Portbou, in a cemetery high on a cliff overlooking the Mediterranean Sea.

Arendt made it out. She and her husband were allowed to leave France and travel through Spain into Portugal, where they spent three months stranded in Lisbon. They passed the hours reading Benjamin's words aloud to one another, and to the small group of refugees who had gathered there, waiting to sail to safety, on the edge of a crumbling world.

≈

Walter Benjamin, it turns out, wrote occasionally for a German magazine called *Uhu*, which means owl. In middle age, at least, with his dark, intense eyes, round glasses, and unruly hair, he looked a little like one.

≈

It cooled off. The wind is blowing again tonight, bending the bamboo along the fence. With the door open I can hear it from my bed. The wind feels like a message, a reminder that other places exist, and other times. I dropped L. at the airport the other day. With her here the city lost a little of its edge. For a while I could ignore its crackling despair. I could laugh at it, at least. Earlier I walked over to Fremont Street to meet people from the Institute for a goodbye drink. On my way there I passed a construction site on Carson. In the months that I've been here, a half block of condos have gone up. I don't know what was there before, but the apartments are finished now. There's still a fence around the site, and a tattered black plastic tarp affixed to the top of the fence. I had never paid any attention to it before, but every few seconds as I was coming and going the wind picked up the tarp, filling it, making it billow like a torn black sail, a chorus of ghosts, shivering. I had to dodge it as I walked past on the sidewalk.

There's a cricket inside the apartment and I can hear it in here too. It's under the bed, I think, singing, seeking a mate or just keeping itself company, telling the night and its fellow crickets tales of this strange and inhospitable land into which it has somehow wandered. And here I am, propped up on the pillows, singing back.

※

At about midnight last night, twenty missiles were fired at Israeli positions in the Golan Heights. None struck their targets, or caused any casualties, or even successfully landed in the Golan. Israel immediately claimed that the rockets were Iranian and within three hours launched a massive assault on Iranian military targets in Syria. According to the Russian defense ministry, twenty-eight Israeli planes crossed into Syrian airspace, firing more than seventy missiles and killing at least twenty-three people. Russia, which controls the most sophisticated air defense system in Syria, took no steps to intervene. This morning the Israeli defense minister claimed that "nearly all the Iranian infrastructure in Syria" had been destroyed. "I hope we've finished this episode," he said.

On both counts, he was almost certainly too optimistic.

※

Another round of goodbye drinks last night. This time with T.— the other fellow, who had been gone on book tour for most of the term—and M. and D., who work at the Institute. Piecing together what happened is like reconstructing a dream. The pieces don't fit, or they keep slipping away, or seem so improbable that they must have belonged to someone else's dream. It was T.'s idea that we could all go bowling at an alley on the Strip, but it turned out that Anthrax was playing and the bowling alley was closed. Or maybe it was open but you had to buy an Anthrax ticket if you wanted to bowl, so we had dinner instead at a Mexican restaurant overlooking a sea of people shuffling through an outdoor mall and then I think the goal was to play bingo in one of the casinos, but we couldn't figure out where to play it or if it could be played at all, so T. shepherded us along the Strip to see the volcano show at the Mirage. After that we could see the dancing fountains at the Bellagio. T. had seen them several times before. The fountains, she promised, would be amazing.

Mainly I remember the crowds on the sidewalks, packed so close that staying together while walking forward took constant focus and effort. All those bodies, strapped into wheelchairs or swathed in shorts or tight-fitting dresses or bikinis and bunny ears, all of us so grotesquely, densely mortal beneath the infinite promise of the neon lights. We passed a man lying barefoot and unconscious, bleeding all over the concrete from a gash in his foot. Like everyone else, we stopped, wondered if we should do something, walked on. The Falun Gong people were out in force with banners and amplifiers broadcasting something about organ theft in Chinese prisons. A truck kept circling on Las Vegas Boulevard towing a mobile billboard with giant photos of nearly naked blond women and the words CALL 24 HOURS GIRLS DIRECT TO YOU. Outside the moat surrounding the Mirage protesters waved signs reading: LIFE IN A BATHTUB IS NO LIFE AT ALL, and MIRAGE DOLPHINS HAVE NO SHADE. I didn't see the dolphins but right on schedule the fake mountain on the far side of the moat erupted in great jets of flame. Spurting fountains transformed water into foaming lava with the help of orange lights. Hillary Clinton's voice crackled out of the Falun Gong speakers behind me. She was speaking sternly about human rights violations in China. Above the volcano towered the hotel, its top floors covered with an enormous ad for the Mirage's Cirque du Soleil franchise, a single word: LOVE.

We followed T. like ducklings. At some point we drifted into the Venetian but all I remember are long colonnades and the gondolas in the canals between the casino and the street and M., beside me, saying, "This is all Sheldon Adelson's," while I wondered what it's like to come here and land a job as a gondolier and pole tourists around the shallow fountains for eight hours and go home to your roommates, get stoned, and watch TV. T. led us farther down or maybe up the Strip toward the Bellagio. Space seemed to have bent. We had veered off the sidewalk and streamed along like so many minnows through the sparkling lobbies of casino after casino until I had no idea where we were or how far we had come. We got there just in time to hear a few lonely, maudlin country chords blasting through the speakers as the fountains began to leap, hundreds of them all at once shooting thousands or tens of thousands of gallons of water higher in the air than seemed possible, the jets leaping in synch with the music, Lee Greenwood suddenly crooning, "'Cause the flag still stands for freedom and they can't take that away . . ."

The water rose and the water fell, arcing and twisting, lit a gleaming white against the dark, still pool beneath us. M. and D. had wandered off somewhere. Lee Greenwood croaked on: "And I'm proud to be American where at least I know I'm free . . ." I stood beside T. and watched the fountains all spurting up together as the song approached a climax that seemed to never end.

T. shook her head, her eyes wide in awe and horror. "I swear," she said, "last time it was Elton John."

≈

The Hermetic corpus also contains a prophecy. In the text known as "The Lament," Hermes Trismegistus addresses his pupil Asclepius and predicts that one day the gods will abandon Egypt. Foreigners will invade the land and forbid the ancient ways of worship. "O Egypt, Egypt," he proclaims, "nothing will survive save words engraved on stones."

Deprived of all routes to the sacred and "weary of life, men will no longer regard the world as worthy object of their admiration and reverence." All of creation will only be a burden to them, "and thenceforward they will despise and no longer cherish the whole of the universe, incomparable work of God." Then "darkness will be preferred to light . . . the pious man will be thought mad, the impious, wise; the frenzied will be thought brave, the worst criminal a good man."

Perhaps this sounds familiar. It will get worse: "Then the earth will lose its equilibrium, the sea will no longer be navigable, the heaven will no longer be full of stars, the stars will stop their courses in the heaven. Every divine voice will be silenced, and all be silent . . . Such will be the old age of the world."

This passage was read by ancient Christian authors, not without some schadenfreude, as a prophesy of the defeat of paganism, written perhaps in the fourth century A.D., after the Christianization of Egypt—a process that was by no means gentle—had already commenced. But the text is almost certainly older than that, and makes no reference at all to Christianity. Apocalyptic literature, as Anathea Portier Young and other biblical scholars have argued, is a form of resistance literature, a coded attempt to envisage some outside in a political present that has become unbearable, even if it means the death of the known world. "The Lament" more likely reflects the anxieties of Egyptians living centuries earlier, under Ptolemaic rule, fearful that the dominant Hellenistic culture would entirely displace the more venerable, indigenous religion. Read backward from the vantage point of the early twenty-first century, it appears also to be prophesying, with astonishing foresight, the banishment, by eighteenth-century Europeans, of the Hermetic tradition itself, and of nearly all extra-European and vitalistic perspectives on existence until we are left with a dead, mechanical husk of a world, haunted by the occasional Lilith. Like all good prophesies, it is at once vague enough and specific enough that it might apply to almost any era. Giordano Bruno, in the late sixteenth century, was certain it applied to his. Might it not also be talking about us, these rising seas and smoke-clogged skies, the bullets raining down from the casinos, the elevation of the very worst citizen to the highest seat in the land?

It doesn't end there. "When all these things have come to pass," the Lament continues, God "will annihilate all malice." He'll do it the old-fashioned way, with flood or fire or disease. "Then he will bring back the world to its first beauty, so that this world may again be worthy of reverence and admiration . . . That's what the rebirth of the world will be; a renewal of all good things, a holy and most solemn restoration of Nature herself, imposed by force in the course of time."

❧

You may find this hard to believe, but I haven't mentioned even half the owls I've seen. Once you start noticing something, you see it everywhere. I see them on T-shirts, on throw pillows, in framed pastel prints on waiting room walls, on the internet and on TV screens, big-eyed owls hawking allergy pills and travel apps. I see them tattooed on people's bodies all the time. I try not to stare. There's one on a billboard I drive past almost every day, selling insurance or mufflers or chicken wings. There are plastic ones on rooftops all over this town, intended to scare off all actual birds but usually smeared with white streaks of shit, an advertisement for the intelligence of pigeons.

Nothing means what we want it to, or never just that. Nothing stays put. Owls are messengers, sure, and they are the actual message. They announce death and war and maybe also wisdom, and all the repressed chaos of the ages, everything that we thought was dead but that's coming back hard and biting at our asses as we try to get through the days with just a little dignity. Marija Gimbutas would say that if they foretell death or disaster they herald rebirth too, but do they? Or is it all just an endless chain of linkages and connections, nothing ever really dying, a web that spreads on and on, ensnaring everything in its sticky, woven strands?

It's a lot to take in at once, this web, and in our dizziness and fear that its limitlessness adds up to meaninglessness we can't help but hack a story out of it, following a single path from node to node and ignoring and excluding all the other links. That's what we do. That's what I've done. But the stories that have been winning out these last two-hundred-and-change years—that have been erasing all the other possibilities, the other choices that we had—they have led us here, to this particular regime of power and to this too-warm abyss, to these fires and floods and the Rhino pacing on the stage, and to another moment of choice. If we are to survive we will have to remember, if we can, that there are always other paths, and that this regime can be dismantled just as it was built. And that beyond any individual route or routes, there is the map itself, this sprawling connectedness without terminus or border. It tells a different kind of story, and presents a different kind of choice.

Louis Auguste Blanqui was sixty-six years old and imprisoned in the dungeon of the Château du Taureau when he began, in 1871, to write the book that he would title *Eternity by the Stars*. "The universe is infinite in time and space, eternal, boundless and undivided," Blanqui wrote in the book's opening sentence, an extraordinary start for a man whose literary output had hitherto been composed of furious manifestos, pamphlets, and proclamations. Of all the revolutionaries of the nineteenth century, none was as intransigent, determined, and austere in his wrath as Blanqui. The bourgeoisie, the church, and all the institutions of the French state had no enemy more ferocious and committed. His sole faith was in the dynamic and transformative force of revolution. To Blanqui modern Europe was not a glorious and ascendant civilization, but the heir to centuries of oppression, hypocrisy, and corruption. "Let us destroy the old society," he wrote, "we shall find the new one beneath the ruins."

Blanqui would devote the entirety of his adult life to that attempt, taking part in four unsuccessful revolutions and countless other plots, enduring defeat after defeat with a feverish, ascetic rigor. At twenty-two, in the spring of 1827, he was beaten nearly to death at a demonstration and, undiscouraged, was shot in the throat at another protest that fall. He was arrested for the first time the following year, and in the decades that followed would be sentenced more than nine times, once to exile, twice to death, twice to life in prison. All told he would spend nearly thirty-three years in jail.

His confinement in 1871 must have been among the most painful of the lot. On March 18 of that year, one day after his arrest, the workers of Paris rose up and seized the city, forcing the army and the government to flee to Versailles. By March 28, the Commune had been declared and Blanqui elected its president in absentia. The Communards offered to release all of their prisoners—whose number included generals, a senator, and the archbishop of Paris—in exchange for Blanqui's freedom. The government refused. (Walter Benjamin later blamed Germany's Social Democrats for having "almost entirely" erased Blanqui's name from history, "though at the sound of that name the preceding century had quaked.") Blanqui would miss the most radical experiment in working-class democracy of the century, the revolution he had been fighting for all his life. Perhaps, with his leadership, it would have turned out differently, but ten weeks later it was over. By the end of May, the army had reoccupied Paris, arrested more than forty thousand Communards, and executed a still-unknown number—estimates range from just under seven thousand to more than twenty thousand—shooting them in the streets and in the barracks, burying them in mass graves beneath the parks and squares.

Blanqui spent those months locked in the Château de Taureau, the Castle of the Bull, a fortress of stone built on a rocky island at the mouth of the Bay of Morlaix, off the coast of Brittany. With its sheer walls rising from the water on a naked lump of rock, it would make Alcatraz look welcoming. Blanqui was the fort's sole inmate. His guards are said to have had orders to shoot him if they saw him approach the window to his cell. But sitting or lying on the damp, stone floor beneath the low, vaulted ceiling of his cell, knowing that all was once again lost and that most of his comrades had been killed, he would have been able, at least, to see the stars in the sky above, reeling free in the blackness. Out there in the middle of the bay, far from the smoke and haze of factories and cities, they must have been spectacularly bright. Certainly Blanqui could not avoid hearing the relentless crashing of the waves, and feeling in his bones the repetition of the tides.

Robbed of any form of human solidarity, Blanqui wrote about the stars. He considered the brevity of even their existence: "What are those billions of suns that succeed each other throughout time and space? A deluge of sparkles. This rain fertilizes the universe." Looking up, night after night, he had no doubt that he was gazing into an eternity bounded only by the limits of his own senses. "The enigma of the universe is constantly before our thoughts," he wrote. "The human spirit wants to decipher it at all costs." He threw himself into the task with the same ruthless determination that had spurred him all his life. He regarded his approach as strictly logical and scientific. ("The universe lies before us, open to observation and to reason.") Yet he began the text of *Eternity by the Stars* with a notion that he attributed to Pascal, though his own version of it was closer to that of the martyred Hermetic magus Giordano Bruno: that the universe can be best conceived of as a sphere whose center is everywhere and whose surface is nowhere.

The key paradox that Blanqui set out to untangle was that despite the manifest infinity of the cosmos, all matter is composed of a finite number of elements, or "simple bodies," as he called them. Only sixty-four had yet been identified, from hydrogen to cerium, but Blanqui allowed that the knowledge of his day was incomplete and that there could be as many as one hundred. The periodic table now goes up to 118, but even if there were ten thousand elements, they could only be joined in a finite number of combinations and the problem would remain: How to reconcile the finite quantity of material building blocks with the infinite expanse of creation? "Let us not ask: where shall we find enough room for so many worlds?" he wrote. "Instead, let's ask: where shall we find enough worlds for all this room?"

His solution was a dizzying one: The combinations must repeat, and they must repeat infinitely. Everything that exists exists in endless iterations: this galaxy, this solar system, this planet, the chair or bench or sofa on which you're sitting, the room in which you're reading this, the ache in your back, the itch on your nose, you yourself, in this instant and every other one, everything everywhere a hall of mirrors with no end. There is not just this one familiar world, but an infinity of "brother-worlds," some of them identical to this one except for the swing of an individual electron, a breeze that blows here but not there, a drop of rain that didn't fall, a thought you didn't allow yourself to think. There you kissed someone you were too afraid to even talk to here. Somewhere else you blew a stop sign, held your tongue, had one too many drinks. Every choice we make and don't make spawns a fresh new world, and somewhere else in the cosmos, too far for our telescopes to find it, spins another planet on which we took the other path. A planet on which you didn't say the words you've always wished you could take back, on which you didn't leave her, on which they didn't find you sleeping and arrest you in your bed. And each of those worlds and every other one exists not once or a dozen times, but in infinite repetitions. "Wherever it may be," Blanqui wrote, "the road that must bring the existence of our very planet to completion has already been traveled billions of times. The road is nothing but a copy printed in advance by the centuries."

Blanqui's favored metaphor figured this world too as writing, and nature as a book printed in an infinite edition. "There are— strictly speaking—no originals," he warned, since this universe has neither beginning nor end, but if only heuristically, or because the textual analogy held such appeal, he nonetheless described the universe as "divided between *originals* and *copies*," describing the latter also as "*proofs*," as if God were a printer setting "type combinations" in every possible array. As if the cosmos were an immense printshop. Or a library, like one described by Borges, who, despite their antithetical political outlooks, was fascinated by the man he called "the communist Blanqui," and dropped references to him, only some of them cloaked, in several of his stories and essays. Walter Benjamin was also obsessed with Blanqui, but he read him badly, missing the forlorn, infernal optimism of the insurrectionist's vision. He wrote with a masochistic fascination only of "the terrifying features" of Blanqui's imagined multiverse, concluding, "Humanity figures there as damned."

To be sure, the prisoner's vision of this quantitative eternity— worlds piling up in isolation from one another, condemned to repeat themselves without awareness or improvement—was a lonely one, a bad trip that tastes of madness. Blanqui might call them brothers but there could be no fraternity among the planets. Our endless doubles and all the armies of close approximations of ourselves could not help us, nor could we advise them of the paths they must not take. "That is the terrible part!" he wrote. "No one can warn anyone else."

Still, there is a freedom here, and solace of a sort. "Let us not be alarmed at those globes pouring out of the quill by the billions," wrote Blanqui, in his cell, a quill clutched between his fingers. His was a vision of absolute equality, in which no hierarchy, spatial or temporal, could begin to make sense. No world or sun is more central than any other. No power lasts. All domination is local and fleeting, laid low by the breadth of the infinite. All of us, "we are part of the copy." None of this makes us any less free: "All mankind, identical at the time of hatching, follow, each on their own planet, the road laid out by the passions, and each individual's particular influence contributes to designing that road."

Time collapses here. It can have no linear reach. Everything that might ever happen has already happened, and will again, forever. What is left to fear? The concept of progress becomes a very silly joke. Any possible advancement is "locked up on each earth and disappears with it." Instead, "always and everywhere, on the terrestrial camp, the same drama, the same set, on the same narrow stage, a noisy humanity, infatuated by its own greatness, thinking itself to be the universe and inhabiting its prison like an immensity, only to drown soon along with the globe."

Eternity by the Stars was published on February 20, 1872, a century before my birth. By then the U.K.'s carbon dioxide emissions had already risen by a factor of more than ten thousand. Blanqui wrote to his sister a month before the book's publication, asking her to see to it that copies be made available not only to the press but to the members of the National Assembly and of the tribunal that would judge him at Versailles. It would have been an unusual defense. He didn't get a chance to make it. Three days before its publication, the tribunal sentenced Blanqui to be exiled. If his health had not been destroyed by his confinement, he would likely have been transported to the South Pacific penal colony of New Caledonia along with nearly four thousand Communards. Instead, his sentence was commuted, and he was condemned once again to life in prison. He was released nonetheless in 1879 and returned immediately to the struggle, traveling throughout France to demand amnesty for all surviving Communards. He did not write about the stars again, nor of the infinity of worlds. He died two years later, on New Year's Day, of a stroke, and was buried in Père Lachaise Cemetery in Paris, where the bodies of more than one thousand of his fellow revolutionaries had been dumped in a mass grave while he was locked in the Château du Taureau. He would have found comfort in their company.

In this world, that's how it ended. For Blanqui, at least. Theoretical physicists have since envisaged the cosmos in terms not so different from his, as a patchwork multiverse containing all possible worlds infinitely repeated. If they're right, and if Blanqui was right, then among all those worlds surely there is one—there must be—in which humans have, at the brink of the abyss, stepped back and learned to live inside of time, and to hold each other there. To hold tight to everything outside of us, and everything within, to everything above and below. Perhaps it's not this world. But perhaps it is.

❧

Tomorrow I leave Las Vegas. I don't know if it's a hangover from too much proximity to the destructive energies of the Great American Id last night, or if I'm just sick of moving, but all day the city has felt diffused with sadness. Not its usual bright and brutal buzz but something gentler and simply mournful. I spent most of the day cleaning and packing up the apartment, folding clothes, laying aside a box of food, toiletries, clothing for the woman on the mat. Late in the afternoon I had to drive to campus and then across town to run an errand in the southwestern corner of the city. By then the light was soft and the sun low above the mountains. I was way out in the construction zone, on the edge of town where you can watch the city eat the desert. New subdivisions were going up on both sides of the highway, the houses still plywood skeletons, the roads already planted with paloverdes and desert willows, all of them in blossom, yellow and pink. Heading home, I drove through Summerlin, not far from Red Rock Canyon. Petroglyph motifs were pressed into the concrete highway overpasses, bighorn sheep, lizards, and strange, humanoid silhouettes, the mountains looking on from the west with bored and patient dignity. I rounded the bend and merged onto another freeway and once again the city was laid out beneath me. Not yet that golden web—it was light still—but I could see the entire basin, downtown, the Stratosphere, the Strip, the mountains striped with the shifting shadows of the clouds. It didn't look apocalyptic or damned, just sad, and silly, like a joke that might have worked but had been told badly and had only made everyone feel more alone.

On my last morning in Las Vegas I woke to the news that thirty-seven Palestinian protesters had been shot to death in Gaza while the Rhino's new embassy opened in Jerusalem. Ivanka was smiling like a game show host in all the photos, gleaming with smugness and wealth. Sheldon Adelson was there too, glowing with a different sort of light. Less than fifty miles away, in Gaza, near Khan Younis, tens of thousands demonstrated, marching to the fence that imprisoned them, knowing that Israeli snipers would not hesitate to shoot, that in the previous five weeks dozens had already been killed, and thousands injured. By the end of the day sixty-two were dead, eight of them children, and more than twenty-seven hundred wounded. The hospitals in Gaza were overwhelmed by the injured. I saw photos on Twitter of the floors smeared with blood.

I threw myself into packing and cleaning, loaded the car, and gave the apartment a final sweep. Waze claimed there was an accident on the interstate, so instead of heading back the way I had come, I drove straight down Highway 95. It paralleled the Colorado River and passed through the Chemehuevi Valley within a few miles of the reservations assigned to the Mohave and the Chemehuevi, and not far from Parker, Arizona, where George Laird worked as a blacksmith until Carobeth Harrington walked into his shop and redirected the trajectory of his days. I cut over on Highway 62 and about one hundred miles east of Joshua Tree passed a stand of tamarisk trees blooming a bright and disconcerting pink. They're imports, or invaders, if you prefer to think that way, originally from North Africa and the Middle East. Tamarisks are mentioned in the *Epic of Gilgamesh*, in the Bible, and the Koran. When Set, the Egyptian god of the desert, killed his brother Osiris, hid his body in a chest, and tossed the chest into the Nile, it washed up across the sea and a tamarisk grew around it. Eventually Isis, Osiris's wife and sister, found it and, after more travails and some help from Thoth, brought Osiris back to life. I didn't know about the flowers, that they were such a shocking pink.

Already on that highway Las Vegas felt like a dream, fleeing from my memory, growing fuzzier and more unreal with every passing mile of creosote basin rimmed by jagged hills. Will what we call civilization go like that too, a brutal, gleaming, plasticized absurdity that we will recall less with nostalgia than with befuddlement and wonder that a whole species could consent to live that way? There are other ways. It's not too late to find them. One way or another, we will have to.

I arrived at my friends' house in time for dinner. They drove off to L.A. after we ate and I moved a chair into the driveway and sat outside to watch the sun set. The jackrabbits hopped past me, a few feet away, pausing to scratch themselves and nibble the flowers off a senna bush. There were no clouds, so the sunset wasn't the operatic, god-lit sort that you see in the winter and fall. The sky just slowly changed color, from blue through several shades of yellow to orange and other colors that I cannot name to a darker blue at last, and then to darkness. Venus blinked on as soon as the sun was gone and for a while it was the only light in the sky. The birds went quiet and the bats came out, diving above the creosote. I stayed there until the stars appeared. It took a while. Jupiter showed in the east, chasing Venus across the ecliptic. It seemed like a long time that the rest of the sky stayed blank, a flat blue unbroken except for those two smeary planets until finally Castor and Pollux appeared above the western horizon and then not one by one but all at once they came.

EPI LOGUE

LAND ERS

Yet is there hope. Time and tide flow wide.

—HERMAN MELVILLE

A year passed before I could get back to the desert. Most of a year, eleven months or thereabouts. Eleven full moons, half moons, new moons, etc., eleven swings of one satellite around another. In all that time I can't say that I saw the moon that much or knew when it was waxing or waning. After Las Vegas I moved in with L. in a city far away, in another country on another continent, where people had their own dramas and no one cared about the Rhino or talked about him much except to snort with a sort of bewildered contempt whenever his name came up, as if he were some exotic and comically disfiguring skin disease, unlikely to afflict anyone they knew, too distant and too disgusting to dwell on for long.

❧

I didn't miss him, but I did miss the desert, and the stars. Sometimes at night L. and I would climb the stairs to the roof of our apartment building, hold each other like we used to outside the house in Joshua Tree, and stare up at the sky, but the lights of the city were far too bright, and even on the clearest nights we could make out only a few dim constellations. Once we rented a car and drove a few hours out of town to spend a weekend in what was supposed to be the darkest spot in the country and one of the best places in the whole continent to see the night sky. All the villages there were dying: the young people had moved to the cities in search of work, leaving most of the buildings empty and dark. Both nights we were there turned out to be cloudy. We saw a few occasional shreds of open sky, the blackness of space rich and dense and thick with stars gleaming for a minute or two before the clouds drifted closed, hiding them again.

✍

I kept up with the news at home, the weather, all of it, perhaps even more attentively than when I was living there. The Rhino blundered on, destroying everything that came close to him, but he didn't kill us all and hasn't yet. Not all of us anyway. I don't find that comforting.

The summer was hot, hot here and hotter than ever almost everywhere. I was glad not to be in Las Vegas. In the first week of July it hit the midnineties in northern Siberia. By the middle of that month wildfires were burning north of the Arctic Circle. There were fires in the moors of northern England too, and in Siberia, Greece, Ukraine, and much of the American West. In November fires erased the town of Paradise, California, killing eighty-eight people, more than tripling the grim record for fatalities set by the Oakland hills fire that I survived seventeen years earlier. If Blanqui is right I'm still running from it, my scrappy, nineteen-year-old self racing down that burning hillside in universe after universe. Somewhere else out there I don't make it. Somewhere else California never burns.

*

In this universe the permafrost in parts of Siberia is no longer freezing. Which means it is not permafrost anymore, but mud. In this universe the United Nations' Intergovernmental Panel on Climate Change reported in vertigo-inducing techno-bureaucratic language that if we want to keep global warming below 1.5 degrees centigrade and prevent absolute fucking planetary catastrophe we have only a few years, until 2030, to cut global carbon emissions by 45 percent, and then twenty years to reduce them to zero. Even the dullest experts agree that this will require collective effort of a sort that our species has never yet attempted. Despite all this, carbon emissions are still going up. They're breaking records, in fact.

*

Late in the summer, L. and I took a train to Portbou, the seaside town in northern Spain where Walter Benjamin ended his life. He had arrived there around the same time of year that we did, in September, carrying seventy dollars in American currency and five hundred francs, plus his watch, glasses, passport, pipe, and the vial of morphine tablets—enough for an overdose—that he had carried with him for more than seven years, since the burning of the Reichstag. He had a black valise too, stuffed with papers, among them a manuscript, now lost. His heart was bad, and though a guide had escorted him and a few others through the mountains to the border from France, the terrain was rough and Benjamin was ill. Too weak to continue, he had spent the night outdoors and slept, if at all, exposed in the rocky hills.

Travel within the European Union is these days supposedly borderless, but the police asked L. and me for our passports coming and going, on the platform and on the train. They barely gave ours a glance, but they lingered with great interest over the papers of another traveler, Afghan or Pakistani, who stood a few feet from us beneath the vaulted ceiling of the station. In September 1940, the police in Portbou had showed Benjamin neither hospitality nor mercy. He had a Spanish visa—and another one for the United States, if only he could get that far—but he had no French exit stamp, so the police told him he would be returned to France. The photographer Henny Gurland, who had crossed the mountains with him, recalled: "For an hour, four women and the three of us sat before the officials crying, begging, and despairing as we showed them our perfectly good papers." The police would not be budged: the travelers could spend the night under guard in the Hotel de Francia. Gendarmes would come for them in the morning.

Benjamin understood this as a death sentence. At some point that evening, another of his traveling companions checked on him in his room and found him "in a desolate state of mind." He was lying in bed, she recalled, half-dressed, and "observing the time constantly," staring at the pocket watch that he had laid open beside him. Before he lost consciousness the next morning, Benjamin gave Gurland a card on which he had scribbled a few sentences. She destroyed it and committed his words to memory: "With no way out, I have no other choice but to end it."

Portbou is small and sleepy and I imagine we saw most of it. It was gray that day and the sea was dark and still. We climbed up to the cemetery on a cliff overlooking the water just outside of town and found Benjamin's grave there, or at least the stone that had been erected in his memory a half century after his death. At first they had given him a niche of his own, but after five years what was left of the money he had carried with him ran out—the hotelier had taken his piece, billing for an imaginary four-night stay plus five lemonades, and there were fees charged by the doctor who had declared him dead, the coffin maker, a priest, a judge. Even for the dead a private room costs money in this world, so Benjamin's body, or what was left of it, was transferred to a common grave.

Standing on the rocky path behind the cemetery, looking out over the dull, flat sea, I was overcome suddenly by anger, furious that Benjamin had to stay forever in this shitty little town that had treated him so poorly. He had arrived ill, exhausted, and afraid, and they had found no place for him, no pity and no kindness. They would have done the same to me or to you or to any of us, and their offspring are everywhere these days, on this continent and on mine, cruel and craven, living smugly in their fears. The sun came out and gleamed off the water, blinding me, and then it went away again.

Back in town we looked for the Hotel de Francia. A waiter in a café pointed to the empty lot across the street. The hotel had been there, he told me. It must have been torn down years before, because there were trees growing in the rubble, a young fig and a sycamore among the high grasses. Cheery murals of undersea scenes had been painted on the fence surrounding it. But the waiter was lying. Or he had misunderstood me. Or I misunderstood him: the hotel is still standing. Later I found pictures of it online accompanying a blog post that was only a year old, too recent for a tree to grow. How could I have missed it? In the photos, I could read the stone plaque affixed to the wall, engraved with a sentence in Spanish: "In this house lived and died Walter Benjamin." Beneath it was a quote from Benjamin in Catalan, apparently the most anodyne one they could find: "All human knowledge takes the form of interpretation."

≈

The funny thing is that they got the quote wrong, and that the idea behind it, generic as it is, was not even Benjamin's. The line appears in a letter Benjamin wrote seventeen years before his death to a friend and mentor, the writer Florens Christian Rang. It is a typically dense and theoretical Benjaminian missive about "the task of interpreting works of art." He concludes by apologizing if he is not being sufficiently clear: "Your basic concept truly got through to me. In the final analysis it is manifested for me in your insight that all human knowledge, if it can be justified, must take on no other form than interpretation ..."

I doubt he would have laughed, but I am sure Benjamin would have been able to nod at least at the irony, however grim, that he, who had written so urgently of the need for the historian to be convinced that "*even the dead* will not be safe from the enemy if he is victorious," would be memorialized in the town that killed him with words that were not his.

In June a man entered the offices of a newspaper in Annapolis, Maryland, and shot seven employees, killing five. In August a man walked into a video game competition in Jacksonville, Florida, and shot fourteen people, three of whom died. In September a man entered a skyscraper in Cincinnati and shot six people, killing four of them. Also that month a woman shot seven people in a Maryland drugstore warehouse, four of them fatally. In October a man entered the Tree of Life synagogue in Pittsburgh and shot seventeen people. Eleven of them died. In November a man entered a yoga studio in Tallahassee and shot seven people, killing two of them and then himself. Five days later a man entered a crowded bar in Thousand Oaks, California, just outside L.A., and shot twelve people, all but one of them fatally, before killing himself. Twelve days passed and a man entered a hospital in Chicago and shot three people, killing all of them before being killed himself. In January a man shot five women in a Sebring, Florida, bank, killing all of them. In February a man shot and killed five employees at a boiler manufacturing company in Aurora, Illinois, and then shot and injured five police officers before being killed in turn.

In March I flew to California. All winter it had rained and rained. It had only stopped a few days before I got there. Even from the window of the airplane, the greenness was so bright it made my heart leap. The delta where the San Joaquin and Sacramento Rivers conjoin was largely flooded. I drove with my nieces down along the Sacramento and the river was higher than I had ever seen it, almost overflowing the levees all along the way. From the freeway you could see the marshy lands between Sacramento and Davis submerged for miles in all directions, water covering the lower branches of the trees. It looked like a dream.

I caught myself thinking again about George MacDonald's *Lilith*, which I had read while I was still in Las Vegas. It is among the strangest novels I have ever come across, and among the more wonderful, but I hadn't planned to write about it. It was so different than all the other takes on the Lilith myth that I didn't know what to do with it, or how to think about it yet. MacDonald, who was Scottish, wrote *Lilith* toward the end of his life, in 1895, four years after the death of his favorite daughter. In the end he would outlive four of his children, plus his wife and at least one of their grandchildren. The novel's milieu, for all its whimsy, is one of overwhelming loss.

The plot begins in earnest when the narrator, allegorically named Vane, steps through a mirror in the attic library of his family's ancient home. He finds himself in a "wild country" in which a raven that speaks to him in riddles will be the least of the marvels that he meets. Walking through a forest, he comes across a beautiful woman, naked, half-starved, and nearly dead. He does his best to nurse her back to health. He warms her by building her a bed of leaves and branches over a flowing hot spring, and feeds her a single grape each day. (He peels it first and removes the seeds.) For all his trouble, he wakes each morning with a "burning thirst" and a wound "like the bite of a leech"—first on his hand, later his arm, then his neck. His patient grows steadily stronger. When she revives enough to speak, she blames his bites on a "great, white leech." She caught it in the act, she assures him, and hurled it into the river while he slept.

She's Lilith, of course, fallen angel and onetime bride of Adam. This time she's a glam-goth heroine: vampirish, magnetic, tormented, and pale, and generally accompanied by a leopardess that she dispatches to suck the blood of children. But MacDonald's Lilith could not be more different from Hugo's or Rossetti's. She is not modernity's despised other, but modernity itself: proud, self-mutilating, and murderous, a prisoner of her own self-regard, convinced in her bondage that she is the very height of freedom. The world over which she reigns is a recognizably capitalist one. The allegory, which would have been easy to decode for readers of MacDonald's day, feels no less current now.

This Lilith is a princess, it turns out. Bulika, the city over which she rules, is rich but hardly enviable. There is no water in Bulika, no flowers, no children, no animals save the princess's cruel feline familiar. Poverty is unknown there. It's illegal, in fact. Deformity and illness are penalized with taxes. The people live off the treasures buried by their ancestors. Everything they need is manufactured for them elsewhere, but they are "good at bargaining and buying, good at selling and cheating." They toss their trash over the battlements and let it pile up outside the city's walls. They hate outsiders and each night expel all strangers from the city.

"But there must be some poor!" insists Vane.

"I suppose there must be," his Bulikan interlocutor responds, "but we never think of such people. When one goes poor, we forget him. That is how we keep rich."

It must have been the butterflies that made me think of *Lilith* again, the butterflies and the rain. In L.A. too it had rained and rained. The air was clean still, the hills everywhere that soft and blazing early green. Everywhere I went I saw the same little orange and black butterflies flying stubbornly northwest, a vast migration of millions upon millions of *Vanessa cardui*, a.k.a. painted ladies, their numbers swelled by the exceptional lushness of the spring. It was like the city and all of us in it were rocks in the bed of a river flowing with painted ladies flitting sometimes up and sometimes down but all heading in the same direction, paralleling the spine of the Sierras and the coast. Thousands flew between the semis clogging the Long Beach Freeway, orange poppies bursting along the exit ramps.

This is the bright side of climate collapse: from now on, the scientists say, long spells of extreme drought will be punctuated by years of unusually heavy rainfall. More rain means more flowers and more grasses and hence abundant food for the caterpillars that become painted ladies. It takes them six generations to make the trip from northern Mexico up to British Columbia and Alaska. Individual butterflies live only a few weeks and breed as they travel, crossing borders without a thought, flitting over walls and checkpoints, flying with an almost inconceivable determination toward a destination that neither they nor their immediate offspring will likely live to see.

❧

In MacDonald's version of the myth, Lilith cannot be repressed or expunged. MacDonald, who was a Congregationalist minister before he started writing novels, was too much of a Christian for that. Even Lilith, he hoped, might be redeemed. And if she might, we might. *Lilith* is his theodicy, his attempt to reconcile God's goodness with the manifest shittiness of creation. For MacDonald evil is not a quality or a substance; it can hold no lasting sway. The "queen of Hell, and mistress of the worlds," as Lilith calls herself, is not irrevocably damned so much as disfigured by her pride. She clings to the fantasy of her own autonomy, perfect and self-wrought. "What I choose to seem to myself makes me what I am," Lilith insists. "My own thought makes me me."

But thought has no such powers. It is the movement of divine love, MacDonald suggests, that makes Lilith what she is: the same rushing, generous force that calls forth all creation. (In the novel it is symbolized by water, the nourishing rain that refuses to fall while Lilith rules Bulika.) You don't have to share his very Christian outlook to agree that what makes us us is not what we imagine and declare ourselves to be, but how we fit into the weave of things: past interwoven with present and with future, great-great-grandparents with children not yet born, with ancient lakes and distant stars. To cut yourself out is to cut yourself off. "Something was gone from her," MacDonald writes of Lilith. "The source of life had withdrawn itself," and yet she survived. "She had killed her life, and was dead—and knew it. She must DEATH IT for ever and ever!"

Or maybe not quite that long. Vane leads an army of innocents—children who won't grow up, "the Little Ones," he calls them, or sometimes "the Lovers"—to overthrow Bulika. They ride mounted on baby elephants, on "diminutive horses" and "little bears." Birds and "great companies of butterflies" form an advance guard above them in the air. When at last they reach the city and enter through the gate, they are terrified. The horses panic, the Little Ones quake. "All," MacDonald writes, "except the bears and butterflies manifested fear."

And look at us down here in Bulika, deathing it all day long, scrambling and afraid, fuming in traffic while the butterflies fly calmly on, knowing they won't make it and flitting onward all the same. Do they know? Does it matter if they do? Will it give too much away if I reveal that in the end Lilith yields and lies down weeping, and as she cries it rains? At first the Little Ones are frightened:

"The sky is falling!" says one.

"The white juice is running out of the princess!" another cries.

Soon the rivers fill and roar to life, running fast not with blood or with tears—the "red juice" or the "white juice," in the argot of the Little Ones—but with pure, clean water, "the juice inside the juice."

≈

L. flew out to join me and we drove together to the desert. We couldn't find a place in Joshua Tree—prices had gone up—so we had rented a house in Landers, twenty-five or so miles to the north and west. When we got there the desert was so green that if you didn't already know where you were you wouldn't think to call it desert. There were more flowers than anyone I spoke to had ever seen. Another bright side to climate catastrophe: all this psychedelic lushness, the green hills smeared with yellow and violet and white, desert dandelion and phacelia and pincushion, as if the buzzards and ravens had gotten bored of the same-old drab Mojave pallet and taken paintbrushes to the whole place.

The terrain in Landers is mainly flat or gently climbing bajada, creosote prairie gridded off in two-and-a-half- or five-acre chain-link lots strewn with decaying cars and other junk, half of the houses losing their battles with the elements. When I'd been there before I'd found it sadder somehow than Joshua Tree, more wind-blasted and exposed. I don't know if four weeks there changed that perception much, but when we arrived the land around the house we rented was thick with yellow poppies and Mojave asters, fiddlenecks, and other flowers that I never figured out the names of. As MacDonald put it, "Nothing in this kingdom was dead; nothing was mere; nothing only a thing."

≈

One cloudy afternoon we climbed an old, steep mining road to the top of Goat Mountain, which is more a lump than a mountain but is high enough that the view from the top was spectacular: ridge after ridge of receding sawtooth peaks to the north and the east, the open desert stretching on for miles. The slopes, usually bare, were carpeted with wildflowers. Climbing up, almost drunk on their brightness, I didn't care if the torrent of color was just another sign that everything is coming apart. I spotted an old metal gate that someone must have labored hard to erect there, to keep other people away from their claims. Miners used to dig for gold on Goat Mountain. For most of the first half of the twentieth century men gave decades of their lives to it, hollowing the earth with tunnels and holes, never taking enough ore out of it to do much more than keep the dream alive. The gate no longer blocked anything. It lay a few yards down the slope, rusted and bent, surrounded now by flowers, little white and gold daisy-like things that I looked for in all the books and websites but couldn't figure out the name of. Broken gate flowers. Dig-all-you-want flowers. Little clever blossoms called the-desert-always-wins.

On the road outside our house we saw the painted ladies flying almost every time we went out—not the heaving river of them that I had seen in L.A., but a trickle still, a shallow, ever-northwest-flowing stream. We hiked a canyon close to Landers where there was still water burbling through the wash—the juice inside the juice!—and flowers everywhere, globe mallows and desert blue-bells, Mojave paintbrush a shade of red that otherwise appears only in hallucinations, desert parsley weird as a sea creature or something recently landed from space. A wildfire had burned through that area thirteen years earlier, scorching tens of thousands of acres. The Joshua trees and junipers were dead but standing, their blackened bark peeled off, branches pointing to the sky. Some of the oaks were coming back, though, saplings shooting up beside the charred cadavers of their elders, growing, it seemed, from the same roots.

ʒ

Did I mention that K. and A. had a baby? They named him T. That's another good way to tell the time. He was four months old when we met him, a handsome fellow with serious eyes, studying everything with an almost eerie equanimity. We went for a walk with the three of them, not far from our old house. The cacti had just begun to bloom. Thousands of fat green caterpillars were climbing the stems of the desert dandelions, munching the petals, leaving low forests of headless stalks behind them. What kind of butterflies would they become? They all seemed to be headed in the same direction, up the wash, toward the spot where we had seen the owls.

ʒ

There were no signs to mark the location of the oldest creosote on earth, no billboards, and no gift shop, but in the end it wasn't hard to find. We drove north from Landers into the wide expanse of the Lucerne Valley. It's mainly creosote out there, creosote and sand, plus distant mountains, enormous sky. Farther west there's a historical marker just off the highway marking the site of "The Last Indian Fight in Southern California"—it was more accurately a massacre— but before we reached it we turned onto an unpaved track strewn with sharp stones. I drove, slowly, until a fence began—two lengths of barbless wire strung between steel posts along one side of the road. Enough, hopefully, to prevent the ATVs from speeding through. We got out of the car and ducked between the wires. Despite the sandiness of the soil, flowers were scattered everywhere up there too, desert marigolds and tiny sand verbena, a delicate, faded pink.

The creosotes grew clumped in loose, uneven circles, most just a few feet in diameter. Sand caught in their roots so that with enough time and wind each ring had been raised, like a platform or proscenium above the desert floor. The most ancient ancestor of each ring would have grown in what was now its empty center. As its own roots and branches died, it would have cloned itself with new sprouts expanding outward so that the circumference of the ring was an index of its age. Or a clock, if you prefer. But the rings weren't really empty: they were a stage for the desert's greatest show, time performing its soliloquy, yowling along with the wind.

It didn't take long to find the big one. It was oblong, more rectangular than round, by my measurement thirty paces long. Someone had piled a few rocks in the middle of the ring to mark it, but if you weren't actively looking it would have been easy to walk past it, just another clump of scraggly desert shrubs. In the mid-1970s, Frank Vasek, a botanist at U.C. Riverside, noticed its unusual size and shape in an aerial photograph. He named it King Clone—why he decided it was male I don't know—and later estimated, extrapolating from radiocarbon data, that it was 11,700 years old. Which makes it more than twice as old as the oldest bristlecone pine and more ancient than all but a very few living things on earth.

That means it first took root sometime around the end of the Pleistocene epoch, when glaciers still covered most of Canada. There were lakes all over the Mojave then, two of them just a few miles to the west. The climate had only recently turned dry enough for creosote to thrive, but already there had been people around for at least a thousand years.

The Akimel O'odham people, who live in what is now Arizona, give the creosote an even longer genealogy. In the beginning, they believe, when Earth Doctor, the creator, floated alone in the darkness and nothing existed save the flowing and folding of the dark, he scraped the sweat and dust from his chest and flattened it in the palm of his hand. With it he sculpted the earth. The very next thing he did—even before he made the sky—was plant a creosote.

Little insects wandering in its branches turned the resin in its leaves into a gum called lac. Earth Doctor sang and gathered the lac and pounded it into shapes, forming the mountains and the hard crust of the earth. Only later did he think to make humans.

I cupped a branch in my hand and blew on it. The leaves let off the same sticky, spicy smell as any other creosote. The smell of rain. A pickup truck sped by on the road beyond the fence, its bed heavy with ATVs, raising a giant cloud of dust as it passed. L. had wandered off. I stepped into the center of the ring. I don't like churches but I felt okay kneeling there in the middle of it, in that empty altar. Like I said, it wasn't empty. I dug my hands into the warm sand, and sat for a while listening to the wind pass through the branches and to the buzz, somewhere, of a single bee.

≈

That evening I read online that a gunman had entered a synagogue in Poway, California, the town in which George and Carobeth Laird spent most of their life together. He killed a sixty-year-old woman and shot and injured a rabbi, one other man, and an eight-year-old girl. The shooter, who was nineteen and had been studying nursing, confessed to having lit a fire a few weeks earlier at a mosque in Escondido, where he and his parents had for years attended church. His family lived near Poway, so he must have driven often between the two towns, just as the Lairds had more than eighty years earlier, and whether he knew it or not he must have often passed the place called Green Valley, where George would park the car on the side of the road, leave Carobeth in

the passenger seat, and walk by himself into the rocks, painted in red
with rectilinear labyrinths that showed no way in or out.

≈

Time loops around. Frank Vasek, the botanist, explained: "The exact point at which a segmenting old individual plant becomes an
incipient clone is problematical."

≈

In the late 1950s, the United States Atomic Energy Commission hired
Dr. Janice C. Beatley to study the effects of radiation on the desert ecosystems that the government had at that point been systematically nuk-
ing for nearly a decade. Shortly before the 1962 Sedan blast at Yucca
Flat, Beatley affixed glass dosimeters to creosote bushes to measure their
exposure to radiation. About two miles from ground zero, she found
creosotes that, having escaped total devastation, were merely "covered
in 'a blanket of radioactive dust.'" When Beatley checked their dosimeters that September, two months after the blast, they recorded radiation
levels of between 3,320 and 5,500 roentgens or, if my math is correct,
several thousand times the amount of background radiation the plants
would have absorbed over the course of a normal year. By winter, their
leaves had turned a "brownish-grey." By summer, they were "completely defoliated." But the creosotes had not died. The following September, one year and two months after the Sedan blast, heavy rains fell
on Yucca Flat. "Abundant sprouts" appeared on the leafless, apparently
lifeless plants. The sprouts grew quickly into branches.

≈

It's hard to find much of anything written about that "Last Indian
Fight" in Lucerne Valley, except that it occurred near a place called
Chimney Rock, just above the shore of an ancient lake that dried up

early in the Holocene epoch. Most accounts do not name the tribes who fought there. None that I found bothered to name or count the dead. It was 1867, barely twenty minutes ago on creosote time. Someone—maybe Serrano, maybe Southern Paiute, which may have meant Chemehuevi, or maybe a group of all three—raided a lumber mill in the mountains and burned it. (Fair enough: the settlers were cutting down the trees.) A posse chased them to the desert floor, where, in the words of one local historian, "most of the savages were pinned down in small bands and destroyed one by one."

That's more or less how it went in California in the years that followed annexation: massacre after massacre of a few dozen or a couple hundred, isolated killings of one or three or eight, almost all of them unmarked by any monument. I found the transcript of an interview with a man named Emmanuel Osague, a descendant of the survivors. He told the stories his uncle had told him, of the cowboys who worked a ranch in the Cajon Pass who would shoot at Indians they saw traveling through; of another ranch near what is now the town of Hesperia, where they displayed the heads of three Paiute boys on poles; and of the battle at Chimney Rock. Many people died there, Osague said, and many of those who did were his relatives.

"It was like the end," he said, "yet our stories still survive."

≈

I read an article in the San Diego papers about a local rock art enthusiast who had photographed boulders in Green Valley that appeared to be unpainted and then subjected the images to digital enhancement. Vivid mazes and sharply crosshatched grids appeared on rock faces that had seemed blank to the naked eye. In some cases the images overlapped, suggesting that someone had painted the rocks and then, after enough time had passed that the original pictographs had faded, someone else had painted over them with new ones, which faded too.

❧

All that month in Landers I wanted to walk again in the wash where more than a year ago I had hiked with K. and A., the one that got this started. For symmetry, if nothing else, and because I wanted to see the owls. (From MacDonald again: "I had only a glimpse of him, but several times felt the cool wafture of his silent wings.") But the weeks flew past, work and deadlines kept me indoors more than I would have liked, and before I knew it the month was nearly gone. A day before I had to leave the desert, I walked with K. and A. again. They had T. with them, so we hiked a different wash, a flatter and more open one that would require less climbing through the rocks.

It was late in the day. The light was soft, the shadows getting longer. We talked about A.'s latest book, which he had just finished, and about mine, which I am finishing right now. I'm sure we talked about the Rhino too, and the Democrats' willful impotence, but mainly we talked about the baby, the caterpillars, and the flowers. It had been hot for a few days earlier in the week, and all the phacelia and desert dandelions had dried out. Overnight, half the color in the desert was gone. We talked about the willows too, and the mystery of their scent. The ones we passed there hadn't even bloomed yet, but if you stood a few yards away from them, in apparently random pockets, the scent was overwhelming, as if it traveled via invisible filaments and spilled out through tears in some otherwise indiscernible web. Lupine was blooming in the wash too, and there were a few primroses left that the caterpillars had missed. T. slept through most of this, but he roused himself to watch the willow branches swaying in the wind.

By the time we got back to the car the sun had set. The sky was violet and nearly dark. We said our goodbyes again. I would come back as soon as I could, I promised, but I knew it would be half a year, if I was lucky. By then T. would be a different person, cackling and grabbing like ten-month-olds do, struggling to stuff the whole world in his mouth to find out what it tastes like. (If children are another way of telling time, a kind of living clock, then what about the rest of us, a little bit older and further along?)

The earth would be a little warmer by then. The glaciers would be thinner, the oceans that much higher. By then another summer would have passed. The flowers and all the spring's lush growth would have dried as crisp as kindling. By then surely some new swatch of the planet's surface would be burning. Elsewhere it would rain and rain. If all clocks strive to represent—in various forms, with hands and digits and shadows and bells—the spinning of the earth on its axis, and calendars in their myriad forms represent the interactions of the earth with the sun and the moon and sometimes the rising and setting of the planets, could it make sense to suggest that the earth is also a clock? Just a bigger one, with a wider grip on time? And if the earth is a clock, why not the other planets? Why not Pluto? And what about the sun and the stars and the galaxies, the black holes at their centers, the clusters of galaxies and the larger bodies that they form, all the swirling stuff of the cosmos—what time does it tell?

I drove north back to Landers, the highway dipping through a deep wash and climbing out steeply again. I tried to keep my eyes on both the road and the stars and I did a bad job of it as usual. Castor and Pollux were to my left, Capella just ahead of me. The highway sloped into another wash and just before it rose again something glided past my windshield. For a moment I thought a wire had been strung low from one side of the road to another and an animal— something too big to be a squirrel or a rat—had raced across it like a tourist on a zip line in Las Vegas. I turned my head just in time to catch a glimpse: white wings spread wide against the night. An owl, flying off into the darkness.

Acknowledgments

All books have many authors. This one could not have been written without the love and collective wisdom of my extended, nonbiological family, some members of which appear in these pages, by initial and without: Anthony McCann, Kirsty Singer, Jacob Forman, Sesshu Foster, Dolores Bravo, Kristen Guzman, Angelo Logan, Cesar Palacios-Guzman-Logan, Adam Goldman, Deirdre English, Roberto Lovato, Olivia and Pearl Cuevas-Carle. In Las Vegas, Tayari Jones and Dawn-Michelle Baude helped keep me at least semi-grounded. So did the staff of the Black Mountain Institute, where I was a Diana L. Bennett fellow. Without BMI's generosity and support, I would not have had the time and freedom to think these thoughts, much less to write them down. Livi Watson read countless drafts and helped shape my ravings into whatever semblance of readable prose they have achieved. Better than that: she showed me the stars and the spaces between them, and opened up the sky.

Thanks as always to Gloria Loomis and Julia Masnik, to everyone at Counterpoint, and especially to Dan Smetanka. Thanks too to the ghosts that swirl through the Biblioteca de Catalunya and the ones that whisper in the deserts. I trust you'll let me know if I got it right this time.

Image Credits

© Peter van Agtmael, Magnum Photos

BEN EHRENREICH writes about climate change for *The Nation*. His work has appeared in *Harper's Magazine*, *The New York Times Magazine*, the *London Review of Books*, and *Los Angeles* magazine. In 2011, he was awarded a National Magazine Award. His last book, *The Way to the Spring: Life and Death in Palestine*, based on his reporting from the West Bank, was one of *The Guardian*'s Best Books of 2016. He is also the author of two novels, *Ether* and *The Suitors*.